ASPECTS OF HOMOGENEOUS CATALYSIS

Volume 7

ASPECTS OF HOMOGENEOUS CATALYSIS

VOLUME 7

The titles published in this series are listed at the end of this volume.

ASPECTS OF HOMOGENEOUS CATALYSIS

A Series of Advances

EDITED BY

RENATO UGO

UNIVERSITÀ DEGLI STUDI, MILANO, ITALY

VOLUME 7

KLUWER ACADEMIC PUBLISHERS
DORDRECHT / BOSTON / LONDON

Library of Congress Cataloging-in-Publication Data

(Revised for volume 7)

Aspects of homogenous catalysis.

   Includes bibliographies and indexes
   1. Catalysis. I. Ugo, Renato.
QD505.A37 1988   541.3'95   87-26660

ISBN 0-7923-0888-3

Published by Kluwer Academic Publishers,
P.O. Box 17, 3300 AA Dordrecht, The Netherlands.

Kluwer Academic Publishers incorporates
the publishing programmes of
D. Reidel, Martinus Nijhoff, Dr W. Junk and MTP Press.

Sold and distributed in the U.S.A. and Canada
by Kluwer Academic Publishers,
101 Philip Drive, Norwell, MA 02061, U.S.A.

In all other countries, sold and distributed
by Kluwer Academic Publishers Group,
P.O. Box 322, 3300 AH Dordrecht, The Netherlands.

Printed on acid-free paper

Printed in the Netherlands

Contents

Large Transition Metal Clusters — Bridges between
Homogeneous and Heterogeneous Catalysts?

G. Schmid

Institut für Anorganische Chemie, Universität Essen

4300 Essen 1, Federal Republik of Germany

I.   Introduction

Clusters: We interpret clusters as an accumulation of interacting atoms, here of metal atoms. The number of atoms forming a cluster is not clearly defined. Normally one speaks of clusters having an ensemble of at least three atoms. But where is the transition from small to large clusters? For this article, large clusters will be defined as metal particles beginning with 25 atoms. Clusters change continously into colloidal metals. Therefore this transition cannot be defined sharply as well. Generally, particles with a diameter larger than ~ 100 Å are regarded as colloids. These again turn continously to small crystallites, as they are often used in heterogeneous catalysis.

Somewhere on this way there is the transition from the molecular to the bulk state, where metallic properties are to be recognized for the first time. Somewhere on this way there is also the transition from homogeneous to heterogeneous catalysis. Complexes and smaller clusters can be used for heterogeneous catalysis also, of course, if they are deposited on a support. Very large clusters and colloids, though still in "solution", offer the vertices, edges and faces as catalytically active sites, normally present on larger particles used in heterogeneous catalysis. There–

1

R. Ugo (ed.), Aspects of Homogeneous Catalysis, Vol. 7, 1–36.

fore, large metal clusters and colloids represent materials, where molecular and bulk properties can be present likewise. This does not exclude that large clusters and colloids, deposited on a support, can be used as "real" heterogeneous catalysts. These species have meta–metallic properties [1] even in solution, but their stabilizing ligand shell, which often supports the solubility, restricts the comparison with naked metal particles. The ligands occupy parts of the surface atoms, which is then blocked for catalytic processes, if the ligand itself is not part of the catalytic action.

The study of the properties of big clusters belongs at present to one of the most fascinating fields in physics and chemistry. The application of clusters in catalysis has been expected euphorically. But, until today only few processes have been known, where big clusters detectably play the role of a catalyst. As recently also has been shown, the hope that by carefully directed use of organometallic complexes the understanding of catalytic reactions in homogeneous catalysis would increase, could not be fulfilled [2]. The situation concerning supported metal particles, having reached an enormous importance as heterogeneous catalysts, is different. The deposition on a support stabilizes the crystallites, thus making a ligand shell unnecessary. However, also in heterogeneous metal catalysis, there is a lack of detailed knowledge about reaction mechanisms and the dependency of the catalytic success e.g. on the particle size and the particle structure. The characterization of those particles on an atomic level, as ever possible, before, during and after a catalytic reaction, belongs to the most difficult problems which should be solved in the near future, if catalysis shall not be exercised any longer by the principle of try and error.

The following discussions shall elucidate the connection between large clusters and small metal particles, with other words, they shall help to understand the transition from the molecule to the bulk, as far as it is possible at present. Herewith we also touch interests of catalytic elementary processes.

II. Large Clusters

The synthesis of large metal clusters can principally succeed on two pathways: a) from complexes and smaller clusters, which are transformed to larger atomic aggregates by thermal treatment or by controlled chemical reactions, and b) by direct buildup from atoms in the gas phase or in solution.

a) From complexes and smaller clusters

The number of large metal clusters, which could be synthesized by chemical methods from complexes is small, whereas for small and medium sized clusters this method is the most successful one. The reason is to be seen in the fact that the genesis of the formation of distinct large, ligand stabilized clusters is so complex that no reactions can be planned using stoichiometric rules. On the contrary, it is left to chance if larger clusters are formed at all.

An elegant approach to many smaller and medium sized transition metal clusters, but also to some large transition metal clusters was found by Fenske et al [3] by the reaction of transition metal halides with silylated compounds of the type $E(SiMe_3)_2$ (E = S, Se, Te) or $ER(SiMe_3)_2$ (E = P, As, Sb). The accelerating power for these reactions might be the elimination of trimethylsilylhalides. So, from $(Ph_3P)_2NiCl_2$ and $Se(SiMe_3)_2$ the structurally characterized cluster $Ni_{34}Se_{22}(PPh_3)_{10}$ is formed [4]. The unusual structure consists of a central, penta—

gonal antiprismatic Ni_{14} unit to which five Ni_4 butterfly—fragments are added. The selenium atoms bridge the faces of the Ni_{34} cluster in a μ_4- and μ_5—manner .

44 metal atoms are arranged in anionic Ni—Pt—clusters of the type $[Ni_{38}Pt_6(CO)_{48}H_{6-n}]^{n-}$, built up from a low—nuclear cluster and a simple complex [5]:

$$7[Ni_6(CO)_{12}]^{2-} + 6\ Pt^{2+} \longrightarrow [Ni_{38}Pt_6(CO)_{48}]^{6-} + 2\ Ni^{2+} + 2\ Ni(CO)_4 + 28\ CO$$
$$(n = 6)$$

The proton free complex exists in pH—dependent equilibria with the protonated forms (n = 3,4,5). As cations may serve $[Ph_4As]^+$ and $[Bu_4N]^+$. In the $Ni_{38}Pt_6$ nucleus the atoms are arranged in a cubic close—packed structure like in many bulk metals.

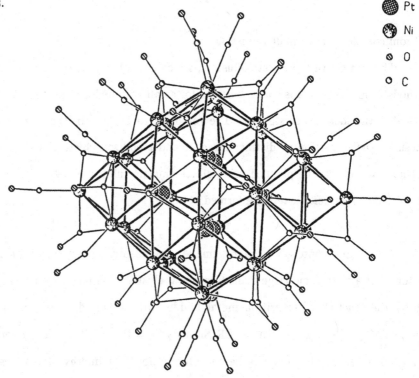

Fig. 1   The molecular structure of the anion $[Ni_{38}Pt_6(CO)_{48}]^{6-}$

Structurally related to this bimetallic system is the anionic, homonuclear cluster $[Pt_{38}(CO)_{44}H_x]^{2-}$ [6], whose synthesis also succeeds by stacking smaller clusters.

A comparatively systematic formation of a large transition metal cluster, beginning with a mononuclear complex, has been described by Gubin [7]. Beginning with PdL_4 via numerous steps, he finally succeeds in $Pd_{38}(CO)_{28}L_{12}$ ($L = PR_3$):

PdL_4 / $Pd(CO)L_3$ / $Pd_3(CO)_3L_4$ / $Pd_4(CO)_5L_4$ / $Pd_{10}(CO)_{12}L_6$ / $Pd_{10}(CO)_{14}L_4$ / $Pd_{23}(CO)_{22}L_{10}$ / $Pd_{38}(CO)_{28}L_{12}$

This reaction sequence is the best characterized stepwise formation of a multi-nuclear cluster at present. The reaction from PdL_4 to $Pd_{38}(CO)_{28}L_{12}$ occurs by Pd^{2+} and CO addition with increasing H^+ concentration.

The formation of another bimetallic cluster, $[(Ph_3P)_{12}Au_{13}Ag_{12}Cl_6]^{m+}$ [8], from Ph_3PAuCl, $(Ph_3P)_4Ag_4Cl_4$ and $NaBH_4$ in ethanol must be of a complex nature and cannot be classified easily. During the simultaneous reduction of two metal complexes, naked gold atoms may form, for the structure shows the cation to consist of a cluster of ten gold and twelve silver atoms forming three interpenetrating triicosahedra. Three gold atoms are positioned in the centres of the icosahedral fragments.

The synthesis of $\{[(p{-}tolyl)_3P]_{12}Au_{18}Ag_{19}Br_{11}\}^{2+}$ from $(p{-}tolyl)_3P$, Au_2O_3, $AgAsF_6$, HBr, and $NaBH_4$ must be of a similar complexibility [9]. The synthesis of these clusters represents the change to the preparation of clusters from metal atoms, described under b). The above mentioned few known examples of large clusters, synthesized from complexes or smaller clusters, show the difficult and accidental approach to these compounds!

b) From atoms

The aggregation of atoms in the gas phase leads to so–called naked clusters with a more or less large size distribution [10]. As naked clusters are chemically very reactive, that means that they aggregate spontaneously to larger units, they cannot be isolated and handled in a chemical sense. Numerous variations of self–made or even commercially available instruments allow the generating of all kinds of clusters from atomic beams. These techniques do not yet play any role in the "molecular chemistry" of clusters, at best for the preparation of metallic particles on a support. They shall therefore not be discussed here. However, it is surprising that by means of the atomic beam technique, it was possible in a few cases "to do chemistry". Recently, the "ligand stabilized" clusters $Fe_{55}(NH_3)_{12}$, $Co_{55}(NH_3)_{12}$, and $Ni_{55}(NH_3)_{12}$ could be detected mass spectroscopically [11]. They have been formed from naked M_{55} (M = metal) clusters in a cluster beam with gaseous NH_3. The number 55 gives us the opportunity to make some remarks to the phenomenon of cluster formation at all, before we shall discuss the synthesis of large clusters from atoms in solution.

The tendency of many metals to form clusters results from the effort to maximize the number of metal–metal interactions in order to maximize its bonding energy, or with other words, to obtain a bonding situation like in the bulk metal. Therefore, it is not surprising that especially the noble metals form clusters very easily on the one hand, and on the other hand the structure of the bulk state is already designed in many large clusters. In smaller clusters, the ligands have a more significant effect on cluster structure. Large clusters are often characterized as wether hexagonal or cubic close packed structures like many bulk metals. Of course, the relationship between ligand stabilized clusters under discussion and the bulk state should not be overvalued. Without doubt, the ligands have a considerable influence on the electronic state in a cluster. Therefore, the properties of ligand stabilized clusters cannot be compared with those of naked clusters. Whereas metallic properties could already be observed for aggregates of 20 − 50 [12] and 60 [13] mercury atoms, respectively, ligand stabilized clusters need probably many more atoms to achieve these properties. This is due to the fact that the interaction of a metal atom with a ligand, e.g. CO, not only consists of a donor bond, but also of a strong π back−bonding, which can even over−compensate the σ donor strength. As a result, there is a loss of electron density on the metal. This could be shown for the group VIII metals experimentally, namely 0.1 e per CO molecule [14,15]. Molecular orbital calculations for small clusters give corresponding values [16]. In the case of the cluster $[Pt_{38}(CO)_{44}H_x]^{2-}$, this leads to an electron transfer of 4 − 5 electrons from the metal nucleus to the ligands [14]. With an increasing number of metal atoms the electronic influence of the ligands on the metal nucleus should decrease. Intensive efforts are therefore focussed on the synthesis of large, ligand stabilized clusters to keep the electronic influence of the ligands as small as possible. In catalysis, those particles should have a special significance, because they possess the electronic properties of a metal, but are still soluble due to their ligand shell: the properties of the bulk metal would be transformed into solutions! New results supporting these ideas will be discussed below.

The relations between complexes, clusters, colloids, and metal crystallites become more understandable following a simple experiment: a metal salt, e.g. $AgNO_3$, is reduced in aqueous solution. A spontaneous precipitation of black silver is observed. Under certain conditions a silver mirror can even be produced. The black precipitate as well as the mirror consist of microcrystalline particles with cubic close–packed silver atoms, aggregated to a polycrystalline material. These complicated processes happen in a few seconds! As we start the reduction with dissolved Ag^+ ions, single Ag atoms must be formed in the frist step. These aggregate to small, but rapidly growing clusters, finally to polycrystalline colloids, which, in special cases, can even be kept in "solution" by protective molecules (see below). Finally, the colloids aggregate to insoluble particles which precipitate. The description of these facts is apparently not attached to any unusual process. But, in reality, it contains a fundamental principle: the generation of metal atoms in solution leads on a direct way to clusters, consisting of close packed atoms! The most common structures of transition metals are the hexagonal (hcp) and the cubic close packed (ccp) ones. Therefore, it is consequent to assume that on the way from the atom to the bulk, clusters with so called magic numbers of atoms exist. Because of their complete outer geometry, these full–shell clusters, consisting of 13 $(1 + 12)$, 55 $(1 + 12 + 42)$, 147 $(1 + 12 + 42 + 92)$ (in general $10n^2 + 2$ atoms in the nth shell) atoms should be long–lived compared to with incomplete shells.

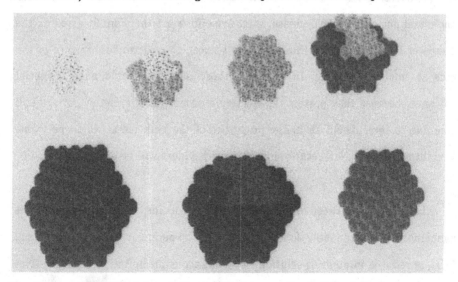

Fig. 2   Simplified description of cluster growth via full–shell clusters

This provides an opportunity to trap just these full–shell clusters chemically by appropriate ligands, present at the right time. Indeed, the interruption of metal formation from atoms in some cases led to defined clusters. But it should be mentioned that metal formation may also be accompanied by processes which are not considered in the simplified presentation in Fig. 2. The fact that the clusters described below are available only in small yields, indicates that the trap of full–shell clusters succeeds only partially and cannot yet be relied on. The following two examples show impressively that large full–shell clusters can be formed from single atoms.

If an acetic acid solution of Pd(II)acetate is reduced by gaseous hydrogen in the presence of small amounts of 1,10–phenanthroline (*phen*), a five–shell cluster consisting of 561 atoms[*] is formed, beside larger amounts of metallic palladium. The compound, isolated by centrifugation from solution, has the idealized stoichiometry $Pd_{561}phen_{36}O_{190-200}$. The oxygen atoms (or molecules) are coordinated to those metal surface atoms that are not covered by *phen* ligands. The oxidation is carried out after the H_2 reduction by contact of the cluster solution with air. So, the cluster becomes air–stable and can easily be handled after its isolation [17]. High resolution electron microscopy, X–ray powder–diffraction, and molecular weight measurements prove the existence of a five–shell cluster with ccp structure, like in the bulk palladium metal.

---

[*]The exact number of metal atoms of course cannot be determined. But as all investigations support the existence of a five–shell cluster, we proceed on the assumption that it consists of the ideal number 561 atoms. Disorders shall lead to deviations in a certain extent.

A cluster of similar composition has been described by Vorgaftik et al [19,20], namely $Pd_{571\pm31}L_{63\pm3}$ (Oac)$_{190\pm10}O_{190\pm10}$, (L = 2,2'–bipyridine, 1,10–phenanthroline). The authors describe its structure to be icosahedral, while the above mentioned Pd_{561} cluster is not, as X–ray powder–diffraction experiments show.

In both cases, a second ligand layer is of interest. The cluster $Pd_{561}phen_{36}O_{190-200}$ is water insoluble after isolation by centrifugation from an aqueous solution. But addition of small amounts of pyridine, acetic acid, propionic acid or any other water soluble aliphatic carbonic acid affects spontaneous dissolution. This phenomenon should be due to a hydrophobic–hydrophobic interaction between the *phen*

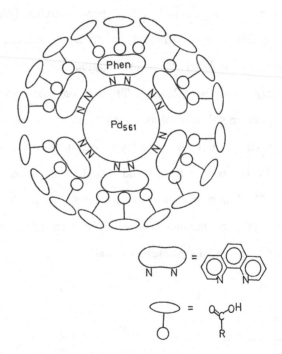

Fig. 3   The simplified conditions around the Pd_{561} nucleus in $Pd_{561}phen_{36}O_{190-200}$ with a second ligand shell of an uncertain number of carboxylic acid molecules.

ligands and the aliphatic moiety of the carbonic acids, and the hydrocarbon—part of the pyridine, respectively. The hydrophilic carboxyl groups and the N atom of the pyridine point outward and cause the water insolubility.

The cluster can also be isolated with its second ligand layer, if it is not centrifuged, but precipitated from solution by alcohol. The presence of acetate in the Vorgaftik cluster is probably to be understood in this sense.

Recently, the synthesis of the first four—shell cluster succeeded: $Pt_{309}phen_{36}^{*}O_{30\pm10}$ [20]. The phenanthroline derivative $phen^{*}$ causes the good water solubility of the compound without any additional ligand shell.

$$Phen^{*} = \quad (\text{structure}) \quad \cdot 2H_2O$$

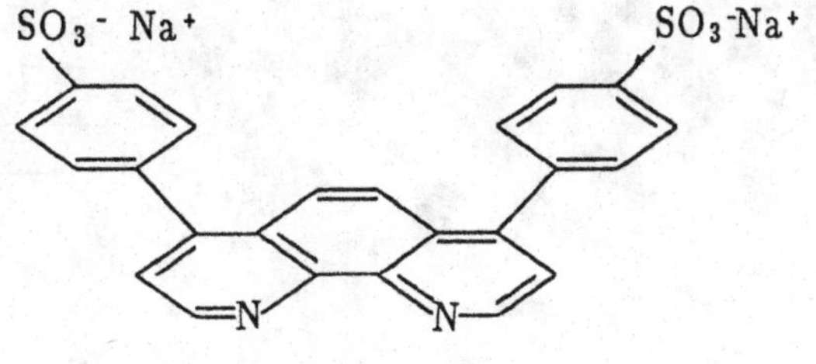

Pt(II)acetate is reduced in an aqueous acetic acid solution by H_2 gas in the presence of appropriate amounts of $phen^{*}$. Free surface atoms are occupied with oxygen from air contact. Then, the black cluster can be handled in air without decomposition. X—ray powder diffraction measurements again show the ccp arrangement of the atoms, like in bulk platinum. The half—width of the X—ray reflections allows the determination of the particle size as well as molecular weight determinations by an analytical ultracentrifuge. The high resolution transmission electron microscopy (HRTEM) not only proves the uniform size of the cluster material, but also provides detailed information on the cluster structure. Fig. 4 shows the HRTEM image of a single Pt$_{309}$ cluster in the [110] direction together with a computer simulated M$_{309}$ particle in the same direction.

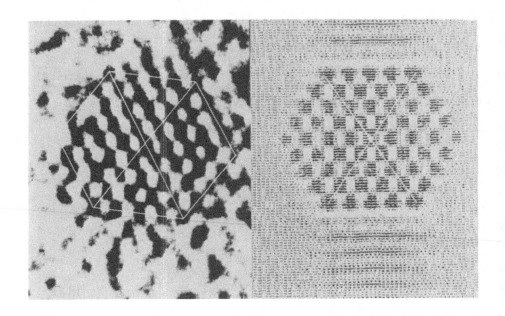

Fig. 4   HRTEM image of a Pt$_{309}$ cluster in the [110] direction (left) and a
              computer simulated M$_{309}$ particle in the same direction (right)

Beside these five– and four–shell clusters a series of compounds has been described, consisting of ligand stabilized two–shell clusters of the ccp type. Table 1 summarizes the known two–shell clusters of the formula M$_{55}$L$_{12}$Cl$_x$.

In contrast to the synthetis of four– and five–shell clusters discussed above, two–shell clusters are prepared by reduction of simple metal complexes like Ph$_3$PAuCl or (Ph$_3$P)$_3$RhCl with diborane. In addition to its reducing power, diborane serves as a Lewis acid to bind excessive phosphine or arsine ligands. A low ligand concentration is important for cluster growth to disired sizes. The cluster $Au_{55}[PPh_2C_6H_4SO_3Na \cdot 2H_2O]_{12}Cl_6$ is prepared by the exchange of PPh_3 in $Au_{55}(PPh_3)_{12}Cl_6$ by the sulfonated derivative.

Table 1   Two–shell clusters of the formula $M_{55}L_{12}Cl_x$

Metal	L		x
Ru	$P(t–Bu)_3$	[21]	20
Co	PMe_3	[22]	20
Rh	PPh_3	[21]	6
	$P(t–Bu)_3$	[23]	20
Pt	$As(t–Bu)_3$	[21]	20
Au	PPh_3	[24]	6
	$P(p–tolyl)_3$	[22]	6
	$PPh_2C_6H_4SO_3Na \cdot 2H_2O$	[25]	6

Due to that ligand, the cluster, otherwise only soluble in dichloromethane or pyridine, becomes water soluble. This opens a completely different kind of chemistry, which might be of special interest concerning catalytic reactions.

A particular property of these clusters helps to understand metal formation a little more: for the first time, the behavior of naked clusters can be studied under mild conditions and not only in the cluster beam. Dichloromethane solutions of these two–shell clusters can be used to generate naked M_{13} clusters (exception: $Au_{55}(Ph_2PC_6H_4SO_3Na \cdot 2H_2O)_{12}Cl_6$) by electrolytical degradation. 20 V d.c., applied to platinum electrodes, dipping into the solutions, cause a cluster–

degradation in a way that the ligands, together with the outer metal atom shell, are lost to give a mixture of different complexes and uncoordinated M_{13} clusters [17,26]. The mechanism of this surprising reaction is unclear. As the degradation happens at the cathode as well as at the anode, indicating that the process is not actually an electrolysis in the classical sense, one can assume that an electron transport inside the cluster nucleus is caused by the contact with the electrodes, leading to the elimination of complex fragments like $AuCl$, Ph_3PAu etc. Finally the meta–stable ccp packed M_{13} clusters result. Indicated by X–ray diffraction investigations, these M_{13} clusters aggregate to superclusters (clusters of clusters), maintaining their structure. In these superclusters, the M_{13} building blocks are linked together via triangle faces in a way that a pseudo close–packed structure is formed, with M_{13} particles instead of atoms. Schematically that process is shown in a simplified manner in Fig. 5.

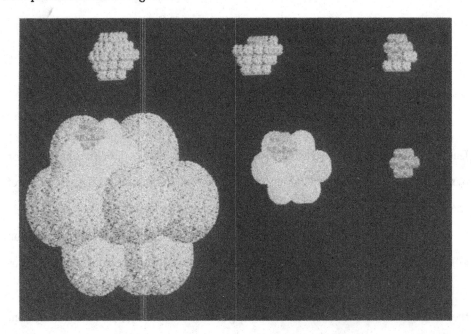

Fig. 5   Schematic representation of supercluster–formation from $M_{55}L_{12}Cl_x$ molecules. The ligands are omitted.

Here, we observe in a controlled experiment what happens during the formation of metal particles from atoms, like in the above described reduction of silver ions to silver metal. There, we supposed the formation of clusters, of course larger ones than here, which aggregate to a polycrystalline material. The well defined super-structures, shown in Fig. 5, correspond to the undefined polycrystalline state in a macroscopic metal particle! The smallness of the M_{13} building blocks makes the $[(M_{13})_{13}]_n$ structures thermodynamically unstable: they degrade within weeks or months into X—ray amorphous powders.

III   Colloids

We defined metal colloids arbitrarily as particles beginning with about 100 Å in diameter up to a size, which still allows them to be kept in "solution". Since ancient times, one knows the beautiful red gold colloids which serve among others to color glass. Colloids in solution show a varying stability, depending on the protecting ligand shell (compare stabilization of clusters). In glass, gold colloids are fixed in a matrix, unable to move. In real solutions, colloids are protected from coagulation either by charges, effecting repulsion of the particles, or by protecting colloids like peptides [27], macromolecules as polyethyleneimine, polyvinyl-imidazole [28], or polyvinylalcohol [29]. The significance of colloids in the past, at present, and in future is described in a review by J. M. Thomas [30]. The neglect of the colloids as "chemical compounds" during the last decades may be traced back to the fact that they could exclusively be handled in very dilute solutions or in matrices. Only in recent years the interest in colloids increased again, as their practical significance in catalysis, photo catalysis, novel optical and micro techno-logical developments, in biology, as adsorbents, sensors and many others has been recognized. Scientific interest increasingly focusses on colloids and the smaller relatives, the clusters, to answer many fundamental questions concerning metal formation. The better handling of colloids, if possible in pure solid form, could lead to completely novel developments.

Recently, the first stabilization of gold colloids was achieved by simple ligands, known from complex and cluster chemistry: by the reaction of gold colloids in aqueous solution with $Ph_2PC_6H_4SO_3Na$, 18 nm gold colloids could be stabilized and isolated as golden leafs [31]. These ligand stabilized colloids can be redissolved in any concentration in water, forming deep–red solutions. The small size distribution (18 ± 1 nm) and the mainly spheric shape make them indeed to a kind of extremely large, ligand stabilized "clusters". Meanwhile it was shown that similar small distributions for any other colloid size is possible. For this purpose, first the

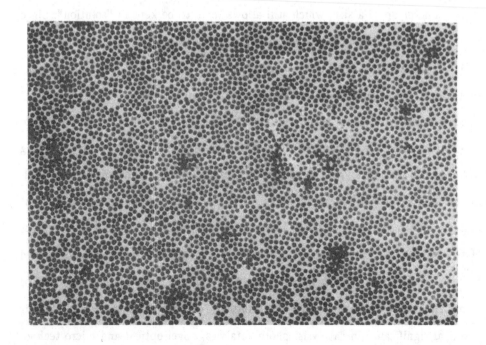

Fig. 6   Phosphine stabilized gold colloids with about 36 nm in diameter

18 nm colloids are generated and after that, goldchloride is added in a quantity, which is necessary to generate a distinct particle size. Then, another reduction process proceeds. The original colloids serve as crystal germs and finally, the phosphine ligands are added for stabilization. The colloids are isolated by precipitation and centrifugation, respectively. In Fig. 6, a kind of monomolecular layer of 36 nm phosphine stabilized gold colloids is shown [22].

Probably for the first time gold colloids could be imaged together with their ligand shell. Such a particle is shown in Fig. 7.

A special property of these ligand stabilized colloids is, among others, that they generate the color of metallic gold on smooth surfaces like glass or paper, if coated by the red solution and dried. The golden color is due to isolated colloidal particles, separated by their ligand shells.

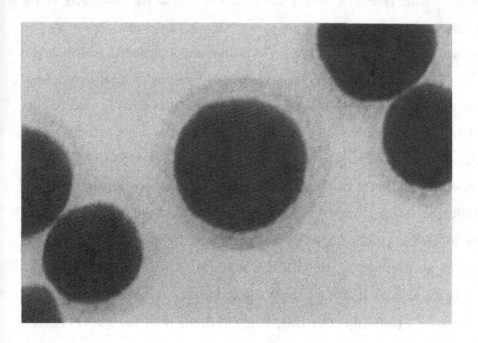

Fig. 7   A gold colloid with its protecting ligand shell

Similar results have been achieved for palladium and platinum colloids, stabilized by the above, but also by other ligands like *phen*[*] or $H_2N–C_6H_4–SO_3Na$. They also produce metallic luster on smooth surfaces. On porous material, the color of the solutions remains unchanged, obviously the separation of the particles is too large.

As follows from numerous electron microscopic investigations, metal colloids often possess different shapes, but they are all built up from crystallites with the atomic structure of the bulk metal.

IV   Supported Metal Particles

Progress in homogeneous catalysis was always linked to developments in organometallic chemistry. Among others, this is due to the fluxionality of the ligands in many organometallic compounds. Ligands in cluster molecules are also characterized by such a high mobility. The fluxionality of chemisorbed molecules on a metal surface is known too and is absolutely necessary for catalytic reactions. For instance, carbon monoxide, chemisorbed on metal surfaces, is electronically related with the covalent bonded carbon monoxide in complexes and clusters. Many investigations have worked at this phenomenon [32–35]. A well analyzed example in the field of small clusters is $Rh_4(CO)_{12}$. It has been investigated by ^{13}C NMR spectroscopy at various temperatures. At $-65°C$, all CO ligands are fixed at the Rh atoms. At $+60°C$ they are involved in such fast migration processes that they all are equalized on the NMR time scale [36]. With increasing cluster size, the ligand mobility in solution seems also to increase, as could be shown for the clusters $Rh_{55}[(t–Bu)_3P]_{12}Cl_{20}$ [23], $Rh_{55}(PPh_3)_{12}Cl_6$ [37], and $Au_{55}(PPh_3)_{12}Cl_6$ [37], respectively. Due to the fast interchange of sites of the phosphine ligands, the expected [103]Rh $-$ [31]P couplings are all averaged so that even at room temperature only one ^{31}P NMR signal can be observed. Migration processes shall run the easier, the more flattened the surface is. The lowest energy for those processes should therefore exist on surfaces of metal crystals. So, the suitable dissociative reactions in organometallic compounds in solution, as well as the high mobility of ligands on surfaces of metal crystallites afford ideal conditions for the homogeneous, and the heterogeneous catalysis, respectively. But, what about clusters in catalysis? So far it has never been shown that a cluster is really a catalyst in a homogeneous catalytic reaction. It may serve as a precursor of the catalytically active species, as degradation processes, leading to different kinds of active fragments, accompany the catalytic event. The only way to go is via supported, and therewith stabilized, clusters, colloids, and metal particles. This fact is recognized more and more, as the number of papers on colloidal systems and supported metal particles increase rapidly.

Back to the metal crystallites: many fundamental investigations on the catalytic properties of metal surfaces have been carried out with model crystals (see for instance [32,38,39]). The results give detailed information about the activities of different faces. A Pt(110) face, for instance, has adsorptive properties, which are different from those of a Pt(111) face. Differences in catalytic activity are also known of terraces, steps, and vertex atoms [40]. This can be understood, as the corresponding atoms have varying coordination numbers and herewith different energies. Fig. 8 elucidates the different coordinated atoms on two faces of a ccp crystal.

Modern techniques of investigation enlarged our knowledge on surface properties enormously: Auger Electron Spectroscopy (AES), UV Photoelectron Spectroscopy (UPS), Low Energy Electron Diffraction (LEED), Secondary Ion– Mass Spectroscopy (SIMS), Extended X–Ray Absorption Fine Structure (EXAFS),

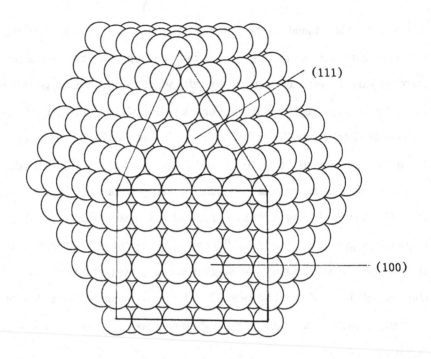

Fig. 8   A cubic close–packed particle of 561 atoms (five–shell cluster) show-
            ing a (111) and a (100) face with the different coordinated atoms.

are only a few of the most important methods. But, the method with the most
informative results is the microscopy. High Resolution Electron Microscopy
(HREM) provided the largest contribution among all methods to the characterization of metal particles. The novel Scanning Tunnel Microscopy (STM) can
complete HREM in the future, first results are promising.

    The study of adsorption processes on metal surfaces requires extremely pure
objects, with other words, the chemistry of surfaces must be carried out under ultra
high vacuum conditions (10^{-9} − 10^{-10} torr). But, under these conditions "normal"
chemistry never runs. Without doubt, we learned a lot about the behavior of surfaces from these fundamental investigations. However, do we therefore know how
this chemistry works? The situation reminds again to the organometallic

chemistry. Many scientists have studied innumerable of so called model reactions and model compounds to understand mechanisms of homogeneous and even heterogeneous catalysis. It shall be reminded to many models for the Fischer–Tropsch synthesis. However, none of these models is really a catalyst and no catalyst can be isolated to be studied. Comparisons with biological processes, working only in living systems, seem obvious: numerous models for nitrogen fixation have been described. None really works under catalytic conditions, like nature does.

A pure metal surface of a (at least almost) perfect crystal is a model. But what happens on a working catalyst with supported metal particles? In addition to the surface structure, attention was payed to the particle size to understand catalytic reactions. In a remarkable paper van Hardeveld and van Montfoort have observed the adsorption of molecular nitrogen on Ni, Pd, and Pt only for particle sizes between ∼ 15 and 70 Å in diameter [41]. This is traced to the fact that particles in this range have a large number of specific surfaces which are not or scarcely observed on smaller or on larger particles. Finally, it is not the size of a particle, but its specific surface that makes it active. This finding is extremely important. The considerations, concerning the adsorbing capacity for nitrogen, are based in the following presumptions: 1. even in very small crystallites (large clusters) the metal atoms occupy crystallographic positions, 2. the particles consist of a constant number of atoms, as they are generated in pores or cavities of the support, so that their growth is limited, and 3. the particles are shaped in such a way that their free energy is a minimum, that means the number of metal–metal interactions is maximized. This holds for spheric particles. If they possess a complete surface, they are cuboctahedra. Uncomplete fcc particles should be numerous, compared with complete ones. They have a larger variety of faces, edges and vertices than those with a complete outer geometry. So, the irregularity of a surface is responsible for the adsorption process, and adsorption is the first step in catalysis.

The most active particle size should be in the range of 18 to 25 Å. Indeed, the strongest nitrogen adsorptions are observed for these particles, which have a maximum of B_5–sites. A B_5–site is defind as a place, where an adsorbed atom touches five others. At complete (111) and (100) faces there are only B_3– and B_4–sites. It seems allowed to transform the results from these investigations to other adsorption and catalytic processes. So, it can be assumed that studies on the size–dependence of crystals give only ostensible results. Actually, the size–dependent surface structure determines the properties. This does not mean that more stable faces cannot be used for adsorption processes. For instance, CO is adsorbed on Pd(100) [42] as well as on Ni(100) [43], Ni(111) [44], or Cu(100) [45]. The comparison of the IR frequences in the ν(CO) region of adsorbed CO molecules with those of metal carbonyls shows that the bonding situations must be very similar: both adsorb between 2150 and 1950 cm^{-1} for terminal coordinated CO, in the region of 1900 – 1750 cm^{-1} for bridging CO groups, and between 1840 and 1600 cm^{-1} for triply coordinated CO ligands. However, such coordinated CO molecules in metal carbonyls are not active catalytically. An excellent review on numerous investigations about adsorption phenomena on defined crystal faces has been written by Davis and Klabunde [46]. We should content with the remark that different surfaces in metal crystallites are different in their activity and that especially terraces, defects, uncomplete surfaces at all in fcc particles are positions of increased activity. The number of these active sites depends on the particle size. The active sites can obviously be compared with the single atoms in a complex in homogeneous catalysis. There is still one important question to be answered: are the identifiable active places in a working heterogeneous metal catalyst identical with the catalytically active sites? Can they explain the fast reactions, which often can be observed? There remain some doubts, if the description of catalytically active supported metals in many papers is correct. Following van Hardeveld and van Montfoort, there is a sufficient large number of B_5–sites only on $18 - 25$ Å particles. However, most of the microscopically identified particles are much larger ($18 - 25$ Å particles are normally too small to be identified). The question is, whether the particles observed and described really are the catalytic active ones. besides the particle size, also the method of preparation influences the adsorptive and catalytic properties [41,47–49], which indicates that the *structure* of the particles is the decisive criterion for catalysis. For heterogeneous metal catalysts with a large size–distribution it is to be supposed that only the particles with an appropriate surface are active, whereas the microscopically well characterized material may be mainly inactive. For instance, if a Pt/C–catalyst with a size–distribution of $20 - 250$ Å is investigated [50], it may be that the most active particles are those with a 20 Å diameter. But, by means of HREM the $100 - 200$ Å species have been characterized.

If the electron microscopic resolution is small, one has to consider another effect. As W. F. Maier recently pointed out [2], heterogeneous metal catalysts often consist of so–called superclusters, that means that an observed particle consists of smaller clusters, as could be shown for the superclusters, composed of M_{13} building blocks [17,26]. Superclusters, observed in Pd catalysts, are put together of about 60 Å building blocks, explaining the catalytic activity [2].

It is not intended in this article to discuss the mechanisms of hetero-geneously catalysed reactions. It should be demonstrated that the supported metal catalysis still is in the state of empiricism. But as shown by numerous investigations, the structure of the surface (dependent on preparation and size) often seems to determine the activity of a metal particle. To understand heterogenous catalysis on a more scientific basis, there must be progress in the characterization of the particles with atomic resolution. Size and preparation are the reason for different surfaces.

A last aspect concerning the action of catalytically active metal particles shall be mentioned, as it is neglected in most of the discussions. In a few investigations of metal clusters and crystallites by means of HRTEM, the dynamic behavior of such particles could be observed with atomic resolution. The knowledge of these results possibly leads to another understanding of heterogeneous reactions in catalysis. As could be shown, gold and platinum clusters in the microscope are subjected to a steady structural rearrangement. Of course, it must be considered that observations in the microscope do not correspond automatically with the catalytic reality. On the other hand, it may be that the high energy in the microscope (up to 400 keV) makes the conditions more realistic than on crystal surfaces under ultra high vacuum and under ambient conditions. The temperature in a particle in a 400 keV microscope is not known, but can be estimated to be a few hundred degrees. So, we are in a realistic temperature range, compared with most of the catalytic conditions. And just here the most remarkable processes on cluster or crystallite surfaces can be observed: outside a 75 Å gold crystallite one can see clouds of atoms which are in a permanent exchange process with certain faces. Indeed, we "observe" the vapor pressure of metal particles [51]. Such clouds of atoms can be seen only outside of (001), (100), (110), and (331) faces, respectively, but never outside the more stable (111) face.

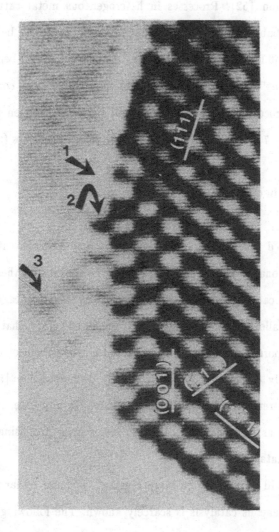

Fig. 10   Clouds of atoms outside a (001) face of a 75 Å gold crystallite

Most of the clouds observed until now reach 7 − 9 Å from the surface into the vacuum. Special activities are also to be observed above vertices. On video recordings it can be followed how an uncomplete vertex of a large gold cluster is in a permanent exchange with atoms in the gas phase, but is not completed during the time of observation [52]. Processes in heterogeneous metal catalysis at higher temperatures probably should follow similar proceedings. It can be concluded that it is not important how many active centres are present at the beginning, but the question is, if the "metal droplet" can form new active sites continously. That these dynamic processes occur easier on small particles than on lager ones, also confirms that 15 − 70 Å particles are more active than larger ones (see above).

V. Large clusters in catalysis

The often assumed function of small clusters in homogeneous catalysis (see e.g. [53,54]) is questionable. Not in any case could it be shown that the starting cluster itself works as a catalyst. That is why the clusters used are called "precursors" [53] and why smaller clusters are used in an "anchored" state, that means they are deposited on supports. In addition, the anchored clusters themselves can be transformed into highly dispersed metal particles by decomposition [53,55]. Little is known about the real active sites and mechanisms. In any case, the use of small anchored clusters as heterogeneous catalysts seems more promising than their use as homogeneous catalysts.

This insight should also apply to large clusters. The use of large clusters ($\geq$25 atoms) in homogeneous catalysis is scarcely known. The following tables 2 and 3 summarize the results of hydrogenation and hydroformylation reactions with $Rh_{55}[P(t–Bu)_3]_{12}Cl_{20}$ in different solvents [56].

Though these results look promising, it must be emphasized that in no case the original cluster could be regained after the reaction. Therefore, $Rh_{55}[P(t–Bu)_3]_{12}Cl_{20}$ must also be regarded as a precursor. The reasons for the degradation of the cluster are understandable. All clusters of this or a similar size

Table 2   Hydrogenation reactions with $Rh_{55}[P(t-Bu)_3]_{12}Cl_{20}$ (50 - 100 mg)

Substrate	Amount	H_2(atm)	Temp.	Time	Solvent	Products	Yields
	44 g	1	0°C	3 h	diacetone-alcohol		50 %
						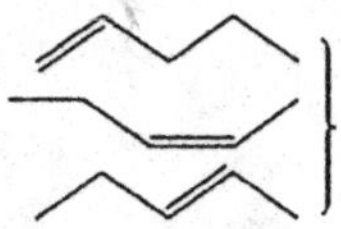	50 %
	44 g	1	RT	3 h	diacetone-alcohol		100 %
	25 g	1	RT	1 d	THF		100 %
	70 atm	80	RT	1 d	diacetone-alcohol		40 %
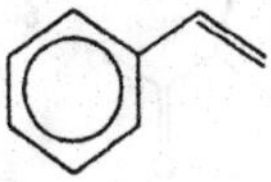	19 g	90	RT	1 d	—	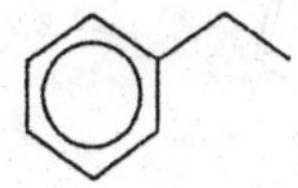	100 %
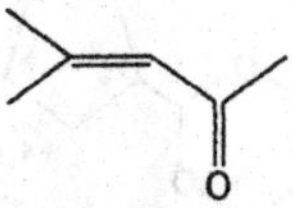	11 g	90	RT	30 h	diacetone-alcohol	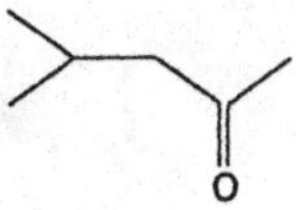	100 %
	18 g	1	RT	4 h	THF	2 Isomers of Hexene	100 %
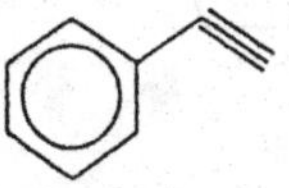	48 g	165	RT	3 d	—	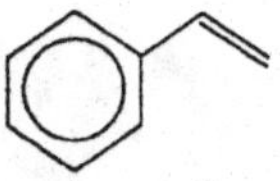	40 %
						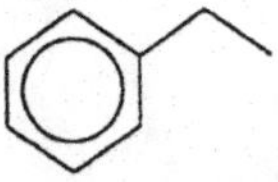	3 %

Table 3 Hydroformylation reactions with $Rh_{55}[P(t-Bu)_3]_{12}Cl_{20}$ (50 – 100 mg)

Substrate	Amount	H_2/CO(atm)	Temp.	Time	Solvent	Products	Yields
	35 g	150	RT	14 d	THF	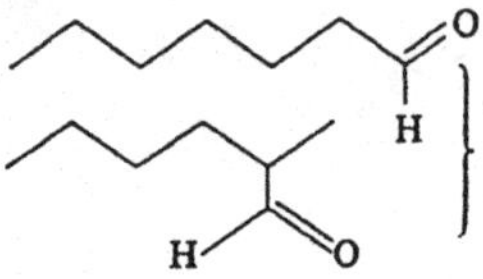	50 %
	20 g	130	RT	4 d	THF	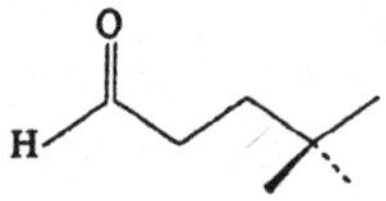	12 %
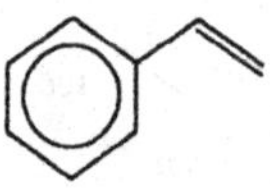	100 g	135	RT	4 d	—	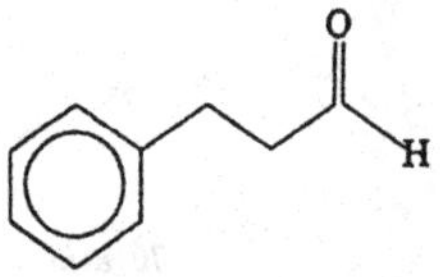	not det
	20 g	200	RT	5 d	diacetone-alcohol	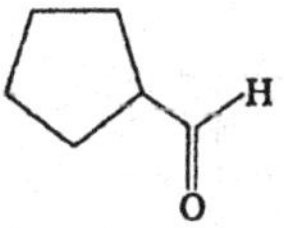	60 %
						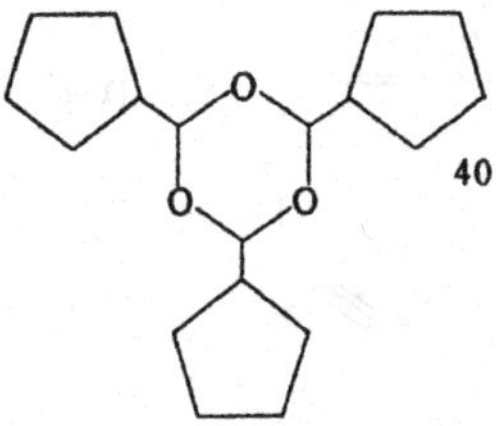	40 %

are unstable in solution, as they are in an equilibrium with dissociated ligands [37], leading to metal formation within a few hours. The future of large clusters in homogeneous catalysis is doubtful. Their structural relationship to complexes and small clusters on the one side, and to metal colloids and particles on the other side cannot be realized chemically. The application of large clusters can only be done via the formation of heterogeneous catalysts. Again there are, like for smaller clusters, two possibilities: a) the anchoring on a support under retention of the ligand shell, or b) the thermal destruction of the ligand shell after deposition on the support to widely uniform metal particles. These two methods promise more success than the homogeneous application, as the clusters are stabilized by anchoring. In addition, naked clusters, generated from ligand stabilized clusters, have the advantage to be unisized, compared with those prepared by common methods.

$Rh_{55}[P(t-Bu)_3]_{12}Cl_{20}$ and $Rh_{55}(PPh_3)_{12}Cl_6$ can be anchored without problems from dichloromethane solutions on Na-Y-zeolite or TiO_2. Hydroformylation reactions of different olefins, catalyzed by such supported clusters is comparable to the best known homogeneous catalysts being in use, but with the advantage of a heterogeneous experimental procedure. This means an easy separation of the products, free of loss, compared with the energy consuming isolation by destillation. In Table 4 the preliminary results of olefin hydroformylation are summarized [57].

Under the reaction temperatures selected, it can be assumed that the clusters still possess their original ligand shell. First attempts with Pd_{561} or Pt_{309} clusters to work for heterogeneous catalysis are promising too. This method of heterogenization of large metal clusters show clear advantages to the homogeneous reactions and possibly also to the common preparation of metal catalysts on a support.

Table 4. **Hydroformylation of different olefins with supported Rh clusters**

Catalyst	Olefin	Temp. (°C)	Pressure (atm)	Products	Turnover rate $\left[\dfrac{M_P}{M_{SA} \cdot h}\right]^{*}$
$Rh_{55}(PPh_3)_{12}Cl_6$ on Na–Y–zeolite	$CH_2=CH_2$	100	300 – 100	H_3CCH_2CHO	>>>10.000
$Rh_{55}[P(t-Bu)_3]_{12}Cl_{20}$ on Na–Y–zeolite	$CH_2=CH_2$	100	300 – 100	H_3CCH_2CHO	>> 10.000
$Pt_{55}[As(t-Bu)_3]_{12}Cl_{20}$ on Na–Y–zeolite	$CH_2=CH_2$	100	300 – 100	H_3CCH_2CHO	≈ 100
$Co_{55}(PMe_3)_{12}Cl_{20}$ on Na–Y–zeolite	$CH_2=CH_2$	100	300 – 100	H_3CCH_2CHO	≈ 20
$Rh_{55}[P(t-Bu)_3]_{12}Cl_{20}$ on Na–Y–zeolite	$H_3CCH=CH_2$	100	300 – 100	$H_3CCH_2CH_2CHO$ 60% $H_3CCH(CHO)CH_3$ 40%	>> 10.000
$Rh_{55}(PPh_3)_{12}Cl_6$ on TiO_2	$H_3C(CH_2)_3CH=CH_2$	100	250 – 200	$H_3C(CH_2)_5CHO$ 42% $H_3C(CH_2)_3CH(CHO)CH_3$ 37% $H_3C(CH_2)_2CH(CHO)CH_2CH_3$ 16%	> 10.000

Catalyst			Substrate	Products	M_P/M_{SA}
Rh$_{55}$(PPh$_3$)$_{12}$Cl$_6$ on TiO$_2$	80	200 – 180	H$_3$C(CH$_2$)$_3$CH=CH$_2$	H$_3$C(CH$_2$)$_5$CHO 48% H$_3$C(CH$_2$)$_3$CH(CHO)CH$_3$ 42% H$_3$C(CH$_2$)$_2$CH(CHO)CH$_2$CH$_3$ 6%	≈ 10.000
Rh$_{55}$(PPh$_3$)$_{12}$Cl$_6$ on TiO$_2$	80	300 – 280	H$_3$C(CH$_2$)$_3$CH=CH$_2$	H$_3$C(CH$_2$)$_5$CHO 53% H$_3$C(CH$_2$)$_3$CH(CHO)CH$_2$CH$_3$ 42% H$_3$C(CH$_2$)$_2$CH(CHO)CH$_2$CH$_3$ 2%	> 10.000
Rh$_{55}$(PPh$_3$)$_{12}$Cl$_6$ on TiO$_2$	125	100 – 50	H$_3$C(CH$_2$)$_3$CH=CH$_2$	H$_3$C(CH$_2$)$_5$CHO 37% H$_3$C(CH$_2$)$_3$CH(CHO)CH$_3$ 33% H$_3$C(CH$_2$)$_2$CH(CHO)CH$_2$CH$_3$ 25%	> 10.000
Rh$_{55}$[P(t–Bu)$_3$]$_{12}$Cl$_{20}$ on Na–Y–zeolite	120	200 – 100	H$_3$C(CH$_2$)$_5$CH=CH$_2$	H$_3$C(CH$_2$)$_7$CHO 43% H$_3$C(CH$_2$)$_5$CH(CHO)CH$_3$ 36% H$_3$C(CH$_2$)$_4$CH(CHO)CH$_2$CH$_3$ 10% H$_3$C(CH$_2$)$_3$CH(CHO)(CH$_2$)CH$_3$ 6%	≈ 8.000
Rh$_{55}$[P(t–Bu)$_3$]$_{12}$Cl$_{20}$ on Na–Y–zeolite	300	300 – 250	H$_3$C(CH$_2$)$_5$CH=CH$_2$	H$_3$C(CH$_2$)$_7$CHO 53% H$_3$C(CH$_2$)$_5$CH(CHO)CH$_3$ 40% H$_3$C(CH$_2$)$_4$CH(CHO)C$_3$H$_2$CH$_3$ 2% H$_3$C(CH$_2$)$_3$CH(CHO)CH$_2$CH$_2$CH$_3$ 1%	≈ 8.000
Rh$_{55}$[P(t–Bu)$_3$]$_{12}$Cl$_{20}$ on Na–Y–zeolite	25	300 – 250	H$_3$C(CH$_2$)$_5$CH=CH$_2$	H$_3$C(CH$_2$)$_7$CHO 29% H$_3$C(CH$_2$)$_5$CH(CHO)CH$_3$ 25%	≈ 1.000

* M_P = Moles of Product, M_{SA} = Moles of Surface–Atoms (42)

VI Conclusions

There is no doubt that metal clusters can be regarded as structural intermediates on the way from complex to bulk metal. This statement is increasingly true with increasing cluster size, as in smaller clusters the structural variety is much more wide–spread. In addition, many physical investigations of large clusters show in some respect a beginning collective behavoir (magnetic measurements, conductivity measurements etc.). The hope that the catalytic–chemical behavior changes in the same manner as the structural properties become more and more metallic, could not be shown in any case. Clusters prove to be good only in the anchored state, with other words in heterogeneous catalysis, since especially large clusters decompose rapidly in solution.

Attempts to use colloidal systems as catalysts in a liquid phase are known [58]. The solutions must be stabilized by polymers. So, the partial hydrogenation of acenapthaline succeeds by means of colloidal palladium:

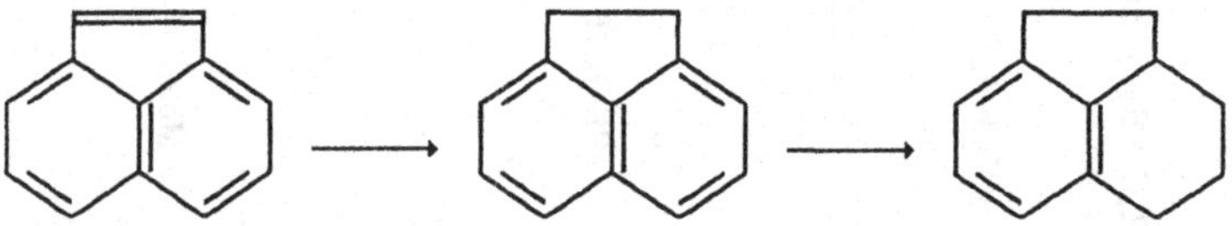

The hydrogenation rate is higher than with palladium black.

The ligand stabilized colloids described in chapter III have not been tested untill now in this respect. However it is to be expected that cluster or colloid catalysts practically will succeed only in the supported state. Such heterogeneous catalysts might have advantages compared with common supported metals. Ligand stabilized clusters possess a uniform size and structure and will such ficilitate any correlation between catalyst structure and catalytic activity. In the anchored state, ligand stabilized and therefrom generated ligand free clusters, represent equivalently working species. First results, as described in chapter IV, are promising and could give novel aspects for heterogeneous catalysis.

References

[1]   M. P. J. van Staveren, H. B. Brom, L. J. de Jongh, G. Schmid, Solid State Commun. **60**, 319 (1986)

[2]   W. F. Maier, Angew. Chem. **101**, 135 (1989); Angew. Chem. Int. Ed. Engl. **28**, 135 (1989)

[3]   D. Fenske, J. Ohmer, J. Hachgenei, K. Merzweiler, Angew. Chem. **100**, 1300 (1988); Angew. Chem. Int. Ed. Engl. **27**, 1277 (1988)

[4]   D. Fenske, J. Ohmer, J. Hachgenei, Angew. Chem. **97**, 993 (1985); Angew. Chem. Int. Ed. Engl. **24**, 993 (1985)

[5]   A. Ceriotti, F. Demartin, G. Longoni, M. Manassero, M. Marchionna, G. Piva, M. Sansoni, Angew. Chem. **97**, 708 (1985); Angew. Chem. Int. Ed. Engl. **24**, 697 (1985)

[6]   P. Chini, J. Organomet. Chem. **200**, 37 (1980)

[7]   S. P. Gubin, J. Chem. Soc. USSR **23**, 3 (1987)

[8]   B. K. Teo, K. Keating, J. Am. Chem. Soc. **106**, 2224 (1984)

[9]   B. K. Teo, M. C. Hong, H. Zhang, D. B. Huang, Angew. Chem. **99**, 943 (1987); Angew. Chem. Int. Ed. Engl. **26**, 897 (1987)

[10] see review E. Schumacher, Chimia **42**, 357 (1988)

[11] A. Masson, Gordon Research Conference, 30.07.–04.08.89, Plymouth, USA

[12] K. Rademann, B. Kaiser, U. Even, F. Hensel, Phys. Rev. Lett. **59**, 2319 (1987)

[13] K. Rademann, Gordon Research Conference, 30.07.–04.08.1989, Plymouth, USA

[14] B. J. Pronk, H. B. Brom, L. J. de Jongh, G. Longoni, A. Ceriotti, Solid State Commun.

[15] B. E. Nieuwenhuys, Surf. Sci. 105, 299 (1981)

[16] C. W. Bauschlichter, J. Chem. Phys. 84, 872 (1986)

[17] G. Schmid, Polyhedron 7, 2331 (1988)

[18] M. N. Vorgaftik, V. P. Zagorodnikov, I. P. Stolyarov, I. I. Moiseev, V. A. Likholobov, D. I. Kochubey, A. L. Churilin, V. I. Zaikovsky, K. I. Zamoraev, G. I. Timofeeva, J. Chem. Soc., Chem. Commun. 1985, 937

[19] M. N. Vorgaftik, J. Chem. Soc. USSR, 23, 37 (1987)

[20] G. Schmid, B. Morun, J.–O. Malm, Angew. Chem. 101, 772 (1989); Angew. Chem. Int. Ed. Engl. 28, 778 (1989)

[21] G. Schmid, W. Huster, Z. Naturforsch. 41B, 1028 (1986)

[22] G. Schmid, unpublished results

[23] G. Schmid, U. Giebel, W. Huster, A. Schwenk, Inorg. Chim. Acta 85, 97 (1984)

[24] G. Schmid, R. Pfeil, R. Boese, F. Bandermann, S. Meyer, H. G. M. Calis, J. W. A. van der Velden, Chem. Ber. 114, 3634 (1981)

[25] G. Schmid, N. Klein, L. Korste, U. Kreibig, D. Schönauer, Polyhedron 7, 605 (1988)

[26] G. Schmid, N. Klein, Angew. Chem. 98, 910 (1986); Angew. Chem. Int. Ed. Engl. 25, 922 (1986)

[27] D. Matuszewska, W. Wojciak, Colloid. Polym. Sci. 255, 492 (1977)

[28] H. Thiele, J. Kowallik, Kolloid. Z. Z. Polym. 234, 1017 (1977)

[29] P. C. Lee, D. Meisel, Chem. Phys. Lett. 99, 262 (1983)

[30] J. M. Thomas, Pure Appl. Chem. 60, 1517 (1988)

[31] G. Schmid, A. Lehnert, Angew. Chem. 101, 772 (1989); Angew. Chem. Int. Ed. Engl. 28, 780 (1989)

[32] E. L. Muetterties, Angew. Chem. **90**, 577 (1978); Angew. Chem. Int. Ed. Engl. **17**, 545 (1978)

[33] E. L. Muetterties, Science **196**, 839 (1977)

[34] H. Conrad, G. Ertl, H. Knözinger, J. Küppers, E. E. Latta, Chem. Phys. Lett. **42**, 115 (1976)

[35] E. W. Plummer, W. R. Salaneck, J. S. Miller, Phys. Rev. **B18**, 1673 (1978)

[36] J. Evans, B. F. G. Johnson, J. Lewis, J. R. Norton, F. A. Cotton, J. Chem. Soc., Chem. Commun. **1973**, 807

[37] G. Schmid, Structure and Bonding **62**, 51 (1985)

[38] G. Ertl, in Studies in Surface Science and Catalysis **29**, 577 (1986)

[39] P. C. Hiemenz, Principles of Colloid and Surface Chemistry, 2. Ed. M. Dekker, Inc., New York, Basel, 1986

[40] G. A. Somorjai, Acc. Chem. Res. **9**, 248 (1976)

[41] R. Van Hardeveld, A. Van Montfoort, Surface Sci. **4**, 396 (1966)

[42] R. J. Behm, K. Christmann, G. Ertl, M. A. Van Hove, J. Chem. Phys. **73**, 2984 (1980)

[43] S. Andersson, J. B. Pendry, Surface Sci. **71**, 75 (1979)

[44] H. Conrad, G. Ertl, J. Küppers, E. E. Latta, Surface Sci. **57**, 475 (1976)

[45] S. Andersson, Solid State Commun **21**, 75 (1977)

[46] S. C. Davis, K. J. Klabunde, Chem. Rev. **82**, 153 (1982)

[47] M. Nakamura, M. Jamada, A. Amons, J. Catal. **39**, 125 (1975)

[48] M. Boudart, A. W. Aldag, L. D. Ptak, J. E. Benson, J. Catal. **11**, 35 (1968)

[49] A. P. Karnaukhov, Kinet. Katal. **12**, 1520 (1971)

[50] J. M. Domiguez, M. J. Yacaman, J. Catal. **64**, 223 (1980)

[51] J. O. Bovin, L. R. Wallenberg, D. J. Smith, Nature **317**, 47 (1985)

[52] J.–O. Bovin, R. L. Wallenberg, private Commun.

[53] B. F. G. Johnson, Transitiom Metal Clusters, Wiley 1980

[54] L. Markó, A. Vizi–Orosz, in: Studies in Surface Science and Catalysis 29, 89
      (1986)

[55] H. Knözinger, in: Studies in Surface Science and Catalysis 29, 121 (1986)

[56] G. Schmid, W. Huster, unpublished results

[57] G. Schmid, B. Küpper, unpublished results

[58] J. S. Bradley, E. Hill, M. E. Leonowicz, H. Witzke, J. Mol. Catal. 41, 59
      (1987)

Transition Metal Catalyzed Synthesis of Organometallic Polymers

Richard M. Laine, Contribution from the Department of Materials Science
and Engineering and, The Polymeric Materials Laboratory of the Washington
Technology Center, University of Washington, Seattle, WA 98195

Abstract: Transition metal catalysts have recently been used to synthesize
organo- metallic oligomers and polymers containing boron or silicon in the
polymer backbone. Three types of transition metal catalyzed reactions have
proven useful for organometallic polymer synthesis: (1) dehydrocoupling by
self- reaction ($2E-H \longrightarrow H_2 + E-E$) or by reaction with an acidic hydrogen ($E-H$
$+ X-H \longrightarrow H_2 + E-X$); (2) redistribution of Si-H bonds with Si-O bonds,
$-[MeHSiO]_x \longrightarrow MeSiH_3 + -[Me(O)_{1.5}]_x-$ and, (3) ring opening polymerization. In
this review, we examine the potential utility of these catalytic methods for
the synthesis of organometallic polymers. In each instance, an effort is made
to illustrate the generality or lack thereof for reaction types. Relevant
literature and proposed reaction mechanisms for each reaction are discussed.

1. **Introduction**

2. **Dehydrocoupling Reactions**

3. **Dehydrocoupling at Boron**

4. **Dehydrocoupling at Silicon**

5. Si-H Self-Reaction Dehydrocoupling

6. Si-H Catalytic Reactions with E-H

7. **Redistribution Reactions**

8. **Ring-Opening Catalysis**

9. **Future Directions**

10. **References**

R. Ugo (ed.), Aspects of Homogeneous Catalysis, Vol. 7, 37–63.

Introduction

The literature contains tens of thousands of publications and patents devoted to the synthesis, characterization and processing of polymers. Despite the fact that there are more than one hundred elements, the majority of these publications and patents concern polymers with carbon backbones. Furthermore, the limited (by comparison) number of publications on polymers that contain elements other than carbon in their backbones are typically devoted to polymers based on silicon, especially those with Si-O bonds.

This disparity is partially a consequence of the dearth of low cost organometallic feedstock chemicals potentially useful for polymer synthesis. It also derives from the lack of general synthetic techniques for the preparation of organometallic polymers. That is, by comparison with the numerous synthetic strategies available for the preparation of organic polymers, there are few such strategies available for synthesizing tractable, organometallic polymers.

In recent years, commerical and military performance requirements have begun to challenge the performance limits of organic polymers. As such, researchers have turned to organometallic polymers as a possible means of exceeding these limits for a wide range of applications that include: (1) microelectronics processing (e.g. photoresists) [1]; (2) light weight batteries (conductors and semi-conductors) [2]; (3) non-linear optical devices [3] and, (4) high temperature structural materials (e.g. ceramic fiber processing) [4,5].

These requirements also challenge the organometallic chemist to develop new, general synthetic techniques for the preparation of organometallic polymers. Unfortunately, preparative methods that have proved exceptionally useful for synthesizing organic polymers are frequently useless or unacceptable when applied to organometallic polymer syntheses. For example, the catalytic synthesis of polyolefins occupies a very large niche in organic polymer synthesis; however, these reactions require element-element double bonds in the feedstock chemicals. Element-element double bonds are relatively rare and difficult to synthesize in organometallic compounds. Consequently, this avenue for the catalytic synthesis of organometallic polymers is not viable.

It is well recognized that the only commercially successful polymers, polysiloxanes and polyphosphazenes, are synthesized by ring-opening polymerization, a process which has analogy in organic chemistry. However, it is our opinion that new, general synthetic routes to organometallic polymers will arise from chemistries with limited or without analogy in organic chemistry. In this review article, we explore the potential utility of catalytic methods of synthesizing organometallic polymers. In particular, we will focus on transition metal catalyzed dehydrocoupling reactions, redistribution reactions and ring-opening reactions.

In each instance, we will illustrate the generality of a particular reaction, discuss the literature and the proposed reaction mechanisms.

Dehydrocoupling Reactions

Dehydrocoupling reactions, as illustrated by reactions (1) and (2), offer great utility for the synthesis of organometallic polymers because H_2 is

$$2E\text{-}H \xrightarrow{\quad catalyst \quad} H_2 + E\text{-}E \tag{1}$$

$$E\text{-}H + X\text{-}H \xrightarrow{\quad catalyst \quad} H_2 + E\text{-}X \tag{2}$$

$$E = \text{Element}, X = O, N, S, E \text{ etc}$$

generated coincident with product. H_2 is easily removed and permits one to drive a thermodynamically unfavorable reaction. Because purification procedures are minimal, the risk of contaminating a reactive and/or thermodynamically unstable polymer is minimized. If the catalyst exhibits sufficiently high activity, then its contribution to the impurity level is also minimized.

Dehydrocoupling at Boron

Transition metal catalyzed dehydrocoupling polymerization has only been observed for boron and silicon compounds. One of the earliest reports of catalytic dehydrocoupling is that Corcoran and Sneddon [6] on the catalytic dehydrocoupling of boron hydrides and carboranes, e.g. reaction (3):

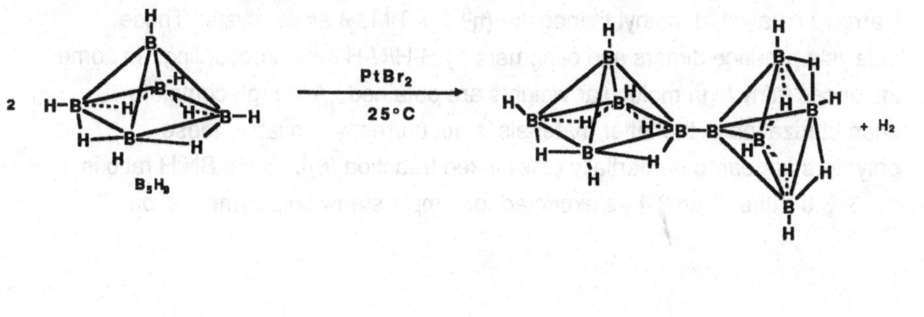

$$2 \quad B_5H_9 \xrightarrow[\quad 25°C \quad]{\quad PtBr_2 \quad} \quad + H_2 \tag{3}$$

Reaction (3) is the only true instance wherein a catalyst has been used to couple B-H bonds to form H_2 and a B-B bond. Unfortunately, this reaction is extremely slow, ≤1-2 turnovers/day/mol cat even at catalyst concentrations of 10-20 mole percent of the borane reactants. Reaction rates do not improve at higher temperatures and the reaction does not lead to true polymeric species. However, recent studies with different B-H containing systems have proven more successful.

Thus, a patent by Blum and Laine [7] briefly describes ruthenium catalyzed dehydrocoupling of N-H bonds with B-H bonds to form H_2 and B-N bonds. Two of the reactions discussed in the patent are:

$$BH_3 \cdot NMe_3 + nPrNH_2 \xrightarrow{\text{Ru}_3\text{(CO)}_{12}/60°C/Benzene} -[nPrNBH]_3^- + H_2 + NMe_3 \qquad (4)$$

$$BH_3 \cdot NMe_3 + MeNH_2 \xrightarrow{\text{Ru}_3\text{(CO)}_{12}/60°C/Benzene} H_2 +$$

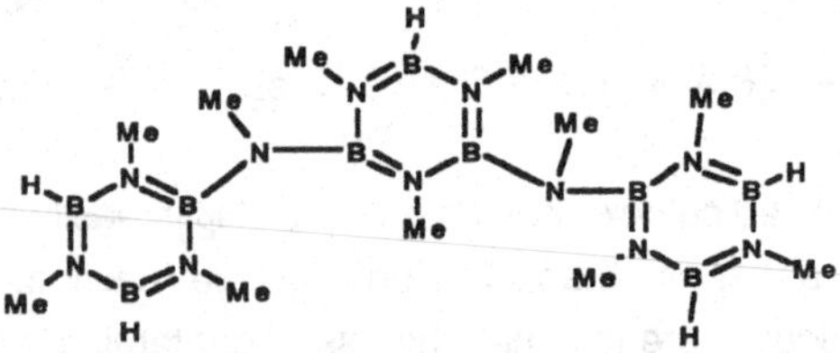

$$(5)$$

Catalyst quantities are normally 0.1mole percent of the borane complex. In the absence of catalyst, reaction (4) gives only the NMe_3 displacement product, $BH_3 \cdot NH_2nPr$. In reaction (5), an oily oligomeric material is recovered in approximately 40% yield. Elemental and mass spectral analyses suggest that the product is the trimeric species shown; however, these results require further clarification.

More recently, Lynch and Sneddon [8,9] have begun to study reactions related to reaction (5) using borazine as substrate and $PtBr_2$, $RhCl(PPh_3)_3$ or Harrod's catalyst, dimethyl titanocene (η^5-Cp_2TiMe_2) as catalysts. These catalysts produce dimers and oligomers by B-H/N-H dehydrocoupling. In some instances fairly high molecular weights are obtained. Although complete characterization of the latter materials is not currently available, these polymers appear to be partially crosslinked [reaction (6)], as the BN:H ratio is ca. 3: 3.8 rather than 3:4 as expected for simple stepwise polymerization.

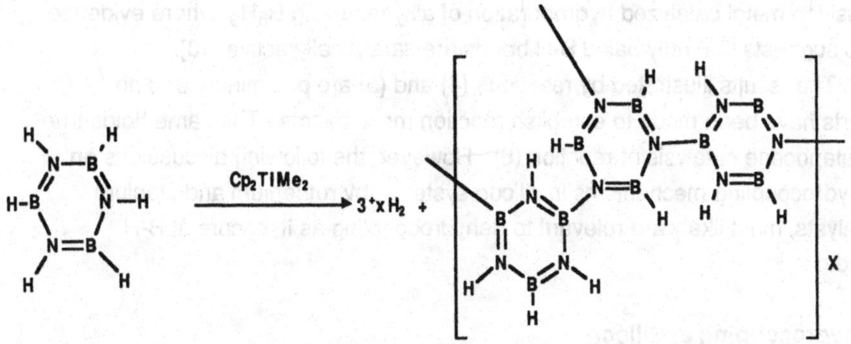

(6)

Corcoran and Sneddon [6] propose a tentative mechanism for PtBr$_2$ catalyzed coupling of B-H bonds as illustrated in Scheme 1:

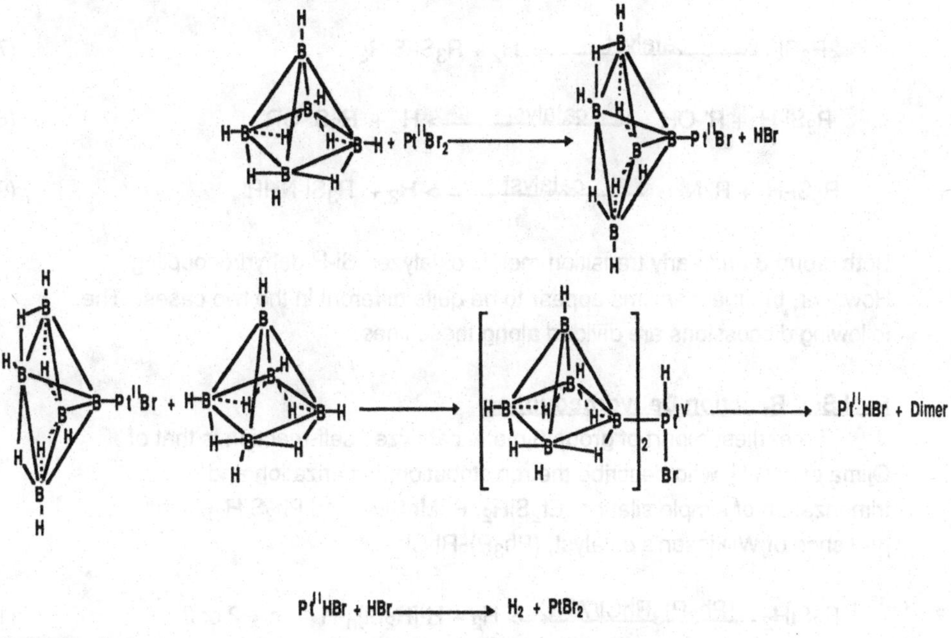

Scheme 1.

If the apical site is blocked with a methyl group, electrophilic substitution does not occur. This, coupled with the exclusive formation of the 1:2'-[B$_5$H$_8$]$_2$ isomer, is consistent with a mechanism where oxidative addition occurs only at a basal site. These results are also consistent with earlier work on transition metal catalyzed hydroboration of alkynes using B_5H_9, where evidence also suggests that only basal B-H bonds are catalytically active [10].

The results illustrated by reactions (4) and (5) are preliminary and no efforts have been made to establish reaction mechanisms. The same holds true for titanocene catalysis of reaction (6). However, the following discussions on dehydrocoupling mechanisms in silicon systems, by ruthenium and titanium catalysts, most likely are relevant to dehydrocoupling as it occurs at B-H bonds.

Dehydrocouping at Silicon

The most well studied dehydrocoupling reactions are those involving Si-H bonds. Si-H dehydrocoupling can take two forms; self-reaction and reaction with acidic E-H bonds:

$$2R_3Si\text{-}H \xrightarrow{\hspace{0.5cm} catalyst \hspace{0.5cm}} H_2 + R_3Si\text{-}SiR_3 \tag{7}$$

$$R_3Si\text{-}H + R'\text{-}OH \xrightarrow{\hspace{0.5cm} catalyst \hspace{0.5cm}} H_2 + R_3Si\text{-}OR' \tag{8}$$

$$R_3Si\text{-}H + R'\text{-}NH_2 \xrightarrow{\hspace{0.5cm} catalyst \hspace{0.5cm}} H_2 + R_3Si\text{-}NHR' \tag{9}$$

Both group 8 and early transition metals catalyzed Si-H dehydrocoupling. However, the mechanisms appear to be quite different in the two cases. The following discussions are divided along these lines.

Si-H Self-Reaction Dehydrocoupling

The earliest report of group 8 metal catalyzed self-reaction is that of Ojima et al [11], who describe the redistribution, dimerization and trimerization of simple silanes (Et_2SiH_2, $PhMeSiH_2$ and Ph_2SiH_2) in the presence of Wilkinson's catalyst, $(Ph_3P)_3RhCl$:

$$R_2SiH_2 \xrightarrow{(Ph_3P)_3RhCl/70°C} H_2 + H\text{-}[R_2Si]_n\text{-}H \quad n = 2 \text{ or } 3 \tag{10}$$

Corey et al [12] have recently assessed reaction (10)'s synthetic utility (where $R_2SiH_2 = Ph_2SiH_2$, 9,10-dihydro-9-silaanthracene, silafluorene or dihydrodibenzosilepin) by examining the effects of variations in the reaction conditions on the rates and yields of products. Their findings indicate that low catalyst concentrations and higher temperatures favor the formation of trimer,

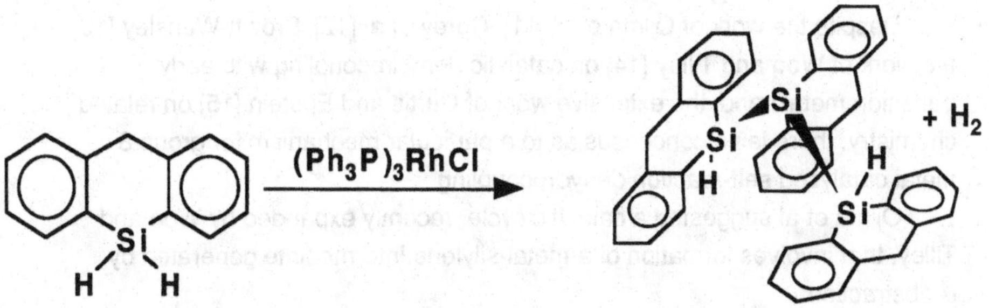

$$+ H_2$$

(11)

reaction (11). The standard reaction uses toluene as solvent, 0.25 mole percent catalyst and reaction temperatures of 20-80°C for 1-48 h. Higher reaction temperatures require less reaction time.

Under identical conditions, the order of reactivities (percent conversion to products) was found to be:

silafluorene (100)> silaanthracene (83)> Ph_2SiH_2 (59) > dibenzosilepin (10)

In a complementary study, Brown-Wensley surveyed the ablity of several group 8 metal catalyst precursors to promote dehydrocoupling of Et_2SiH_2 [13]:

$$Et_2SiH_2 \xrightarrow{\text{catalyst/RT}} H_2 + HEt_2Si\text{-}SiEt_2H \qquad (12)$$

Typical reaction conditions were 1 mole percent catalyst run in neat Et_2SiH_2 for periods of up to 75 h. The following relative activities were observed:

$(Ph_3P)_3RhCl$ (31) > [Pd(allyl))Cl]$_2$ (12) > [Rh(CO)$_2$Cl]$_2$ (5) > $(Ph_3P)_3Pt(C_2H_4)$ (1) ≈ [Rh(COD)Cl]$_2$ (1)> Pt(COD)Cl$_2$ (0.7)> RhCl$_3$ (0.3)> $(\eta^5\text{-}C_5H_5)Rh(C_2H_4)$ (0.2) ≈ [Ir(COD)Cl]$_2$ (0.2) > H_2PtCl_2 (0.1) ≈ $(Ph_3P)_2PtCl_2$ (0.1).

Although some of the catalysts: Pt(COD)Cl$_2$, [Pd(allyl))Cl]$_2$, [Ir(COD)Cl]$_2$ and [Rh(CO)$_2$Cl]$_2$, were observed to dehydrocouple Et_3SiH to $Et_3Si\text{-}SiEt_3$, there was no indication that these same catalysts were able to couple $HEt_2Si\text{-}SiEt_2H$. The hydrosilylation activities of this same set of catalysts gave much the same ordering as their dehydrocoupling activities which suggests that similar initial mechanistic steps are involved in both reactions.

Despite the work of Ojima et al [11], Corey et al [12], Brown-Wensley [13], the work of Woo and Tilley [14] on catalytic dehydrocoupling with early transition metals and, the extensive work of Curtis and Epstein [15] on related chemistry, there is no concensus as to a particular mechanism for group 8 metal catalyzed self-reaction dehydrocoupling.

Ojima et al suggested a catalytic cycle, recently expanded by Woo and Tilley, that involves formation of a metal-silylene intermediate generated by α-abstraction:

$$R_2SiH_2 + M \longrightarrow R_2SiH\text{-}M\text{-}H \tag{12}$$

$$R_2SiH\text{-}M\text{-}H \longrightarrow H_2 + R_2Si{=}M \tag{13}$$

$$R_2Si{=}M + R_2SiH_2 \longrightarrow H(R_2SiH)M{=}SiR_2 \tag{14}$$

$$H(R_2SiH)M{=}SiR_2 \longrightarrow HM\text{-}SiR_2SiR_2H \tag{15}$$

or

$$R_2Si{=}M + R_2SiH_2 \longrightarrow HMSiR_2SiR_2H \tag{16}$$

$$HM\text{-}SiR_2SiR_2H \longrightarrow M + HSiR_2SiR_2H \tag{17}$$

We suggest that reactions (18) and (19) may also be operative for group 8

$$R_2Si{=}M + R_2SiH_2 \longrightarrow R_2HSi\text{-}M\text{-}SiHR_2 \tag{18}$$

$$R_2HSi\text{-}M\text{-}SiHR_2 \longrightarrow M + R_2SiH\text{-}SiHR_2 \tag{19}$$

metal catalyzed dehydrocoupling as discussed below.

The possible participation of a metal-silylene intermediate in the catalytic formation of polysilanes is supported by Zybill et al's report [16] that the HMPT stabilized iron-silylene complex, $(CO)_4Fe{=}SiMe_2$, decomposes readily to polysilanes and $Fe_3(CO)_{12}$. However, Curtis and Epstein argue that it is not necessary to invoke silylene intermediates to obtain plausible explanations for the reactions observed. In fact, silylene intermediates cannot form in the catalytic cycle leading to $Et_3Si\text{-}H$ dehydrocoupling.

If the catalytic cycle follows reactions (12)-(15) and (17), then the complete catalytic cycle would require the metal center to undergo two consecutive oxidative additions and then two consecutive reductive eliminations for a change of four electrons in each direction. Formally, the metal-silylene is formed by oxidative addition of Si-H to the metal followed by α-abstraction. Reaction (14) requires a second oxidative addition. Because it is unlikely that such a catalytic cycle would be observed for all of the metals found capable of promoting dehydrocoupling, it appears that a catalytic cycle involving reactions (16) and (17) and/or (18) and (19) is more reasonable. We believe that the disilyl intermediate shown in reactions (18) and (19) is operative in at least some catalyst systems because it permits us to explain dimerization of R_3SiH without invoking a metal-silylene complex.

Curtis and Epstein note that the nickel triad readily catalyzes H/D exchange at R_3SiH centers. If four electron redox reactions are not operative, then we must consider bimolecular reactions:

$$R_3Si\text{-}M\text{-}H + R'_3Si\text{-}M\text{-}D \longrightarrow MHD + R_3Si\text{-}M\text{-}SiR'_3 \qquad (20)$$

$$MHD + R_3Si\text{-}M\text{-}SiR'_3 \longrightarrow R_3Si\text{-}M\text{-}D + R'_3Si\text{-}M\text{-}H \qquad (21)$$

or

$$R_3Si\text{-}M\text{-}H + R'_3Si\text{-}M\text{-}D \longrightarrow R_3Si\text{-}D + H\text{-}M\text{-}M\text{-}SiR'_3 \qquad (22)$$

$$H\text{-}M\text{-}M\text{-}SiR'_3 \longrightarrow 2M + R'_3Si\text{-}H \qquad (23)$$

Other possibilities also exist. Clearly, the species $R_3Si\text{-}M\text{-}SiR'_3$ could reductively eliminate dimer coincident with MHD eliminating HD. An alternate mechanistic explanation could center on metal clusters as catalytic intermediates. Proof for these possible mechanisms, in the form of kinetic studies, is currently lacking in the literature.

Unfortunately, the group 8 metals have proved useful only for the synthesis of simple oligosilanes not for polysilanes. However, Harrod et al [17-20] have discovered that early transition metals provide quite active catalysts for the synthesis of oligo- and polysilanes. These catalysts hold much promise for the synthesis of true polysilanes via self-reaction dehydrocoupling. Harrod et al first described the catalytic dehydrocoupling of mono-substituted silanes by dimethyl titanocene, Cp_2TiMe_2 or $(\eta^5-C_5H_5)_2TiMe_2$, catalysts in 1985 [17]. Since then, the list of early transition metal catalysts that promote reaction (24) has grown to include; vanadium [18], zirconium [18], and hafnium [14].

$$xRSiH_3 \xrightarrow{\text{catalyst}} xH_2 + -[RSiH]_x- \qquad\qquad (24)$$

Harrod et al also briefly describe the successful use of thorium and uranium catalysts [18].

The oligomers produced from reaction (24), using dimethyltitanocene as catalyst, are atactic, bimodal and hydrogen terminated [19]. Vapor pressure osmometry indicates that the $PhSiH_3$ and n-hexylSiH_3 derived oligomers have molecular weights of 1000-1500 D which corresponds to DPs of about ten and narrow polydispersities. These results are corroborated by size exclusion chromatography studies (GPC) using polystyrene standards. If $(\eta^5-C_5H_5)_2ZrMe_2$ is used, DPs of up to 20 silicon units are observed. The bimodel distribution arises because of the production of cyclic species. The all trans-hexaphenyl-cyclohexasilane crystallizes out of solution if the reactions are left to stand.

Titanocene catalyzed polymerization is extremely susceptible to the steric environment about silicon. Furthermore, hydrogen can compete with silane for the catalytically active site on titanium [19]. Thus, efforts to conduct valid kinetic studies and to quantitatively assess the effects of steric environment about silicon on reaction rate and product selectivity required the addition of a sacrificial alkene, reaction (25). If reaction (25) is run with the following set

$$RSiH_3 + R'CH=CHR' \xrightarrow{Cp_2TiMe_2/toluene/20°C} -[RSiH]_x- + R'CH_2CH_2R' \qquad (25)$$

of silanes, under identical conditions, where R'CH=CHR' = cyclohexene, the relative order of reactivities observed is:

$PhSiH_3$ (13.2) > 4-MePhSiH$_3$ (9.8)> MePhSiH$_2$ (4.6) > PhSiD$_3$ (3.6) > n-hexylSiH$_3$ (1.0) ≈ PhCH$_2$SiH$_3$ (1.0) > c-hexylSiH$_3$ (0.5)

Both c-hexylSiH_3 and MePhSiH$_2$ are unreactive except in the presence of the sacrificial alkene. Under the reaction conditions used, both silanes give only dimer products.

$MeSiH_3$ and SiH_4 were also studied under the same reaction conditions (toluene solvent, 10 mole percent catalyst). Even without the sacrificial alkene, these silanes were too reactive and difficult to work with. Both silanes tend to give intractable, highly crosslinked polymers. These results are indicative of the severe influence of steric effects at silicon and on the efficacy of the catalytic reaction.

Harrod et al have also examined the reactivity of germanes under the same conditions [20] and find that they are actually more reactive than their silicon counterparts. The only previous effort to couple germanes was that of Marchand et al [21], who find discovered reaction (26):

$$Et_2GeH_2 \xrightarrow{\text{steel wool/75°C}} xH_2 + H-[Et_2Ge]_x-H \quad x = 2-4 \qquad (26)$$

Unless care is taken, the monosubstituted germanes also react to give highly crosslinked polymers. Thus, with dimethyltitanocene, $PhGeH_3$ reacts at room temperature to give a gel. However, vanadocene (a poor catalyst for silanes) provides effective stepwise polymerization at 50°C to higher oligomers and polymers that were not further characterized.

Although not useful for monosubstituted germanes, dimethyltitanocene proved to be an effective catalyst for the polymerization of Ph_2GeH_2, a reaction not possible with the analogous silane. This result is extremely important in defining a mechanism for both the silane and germane the polymerization reactions assuming the mechanisms are the same.

Harrod et al [17-20] and Tilley et al [14,22] have undertaken detailed studies to elucidate the reaction mechanisms whereby early transition metals promote self-reaction dehydrocoupling in an effort to develop better catalysts. The long term goal is to catalytically prepare high molecular weight polysilanes and polygermanes with well defined properties.

Harrod et al have isolated the following complexes from reaction solutions:

Unfortunately, these complexes do not appear to be true intermediates in the catalytic cycle. Woo and Tilley [14,22], in an effort to slow the catalytic

reaction down so that intermediates can be isolated, have studied the
zirconium and hafnium systems. They find that the CpCp*ZrClSi(SiMe$_3$)$_3$ and
CpCp*HfClSi(SiMe$_3$)$_3$ (where Cp* = Me$_5$Cp) complexes undergo stoichiometric
reactions with PhSiH$_3$, reaction (27). The product of this stoichiometric
reaction decomposes, reaction (28), to form -[PhHSi]$_x$- polymers rather than
oligomers:

CpCp*MClSi(SiMe$_3$)$_3$ + PhSiH$_3$ $\underline{RT(M = Zr \ or \ Hf)}$>

$$HSiMe_3 + CpCp^*MClSiH_2Ph \qquad (27)$$

CpCp*MClSiH$_2$Ph $\underline{RT(M = Zr \ or \ Hf)}$> -[PhHSi]$_x$- + CpCp*MHCl $\qquad$ (28)

Typically, the presence of chloride substituents limits reactions to
stoichiometric events or totally inhibits reaction. Thus, Cp*ZrSi(SiMe$_3$)$_3$Cl$_2$ is
inactive, while both Cp$_2$Ti(SiMe$_3$)Cl and Cp*Ta(SiMe$_3$)Cl$_3$ will only dimerize
PhSiH$_3$. In contrast, the complex, CpCp*Zr(SiMe$_3$)Si(SiMe$_3$)$_3$, can be used to
successfully catalyze the polymerization of PhSiH$_3$ to give much higher
molecular weight polymers with DPs of up to 40, coincident with the formation
of cyclomers. Likewise, Chang and Corey report that in contrast to titanocene,
which will dimerize MePhSiH$_2$ only in the presence of alkene, zirconocene will
oligomerize MePhSiH$_2$ to give at least pentasilanes [12b].

The proposed reaction mechanism is unlike that suggested above for the
group 8 metal catalysts. It probably requires σ-bond metathesis. These are
reactions which have four-center transition states:

$$L_nM\text{-}R \ + \ H\text{-}R' \longrightarrow \begin{bmatrix} \overset{\delta^-}{R'} \cdots \overset{\delta^+}{H} \\ L_nM \cdots R \\ \underset{\delta^+}{} \quad \underset{\delta^-}{} \end{bmatrix} \longrightarrow L_nM\text{-}R' \ + \ H\text{-}R$$

Reactions (27) and (28) are representative of a σ-bond metathesis
reaction. The activation parameters for reaction (28) where M = Hf are $\Delta H^{\ddagger}$ =
16.4 kcal/mol, $\Delta S^{\ddagger}$ = -27 eu and K_H/K_D = 2.5. The values parallel those found
for previously observed σ-bond metathesis reactions [23].

If we accept this as the likely mode of reaction, then we must still outline a reasonable catalytic cycle. Kinetic studies of the decomposition of CpCp*HfClSiH$_2$Ph to CpCp*MHCl and -[HSiPh]$_x$- reveal a second order rate dependence on [CpCp*HfClSiH$_2$Ph] with $\Delta H^{\ddagger}$ = 19.5 kcal/mol, $\Delta S^{\ddagger}$ = -21 eu. Woo and Tilley suggest that this indicates a four center transition state and permits them to propose a polymerization reaction such as shown in Scheme 2:

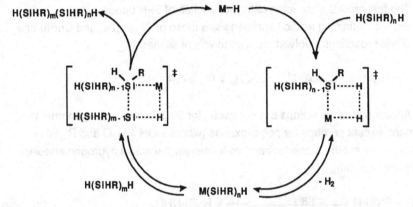

Scheme 2.

Harrod et al have argued [18] that metal-silylene-like intermediates formed by α-abstraction play a role in the early transition metal catalyzed polymerization of silanes; however, their ability to catalyze polymerization of Ph$_2$GeH$_2$ [20] provides evidence against this type of intermediate. Woo and Tilley have also explored the possibility that metal-silylene intermediates might play a role in their polymerization studies. However, when reaction (28) is carried out in the presence of silylene traps such as HSiEt$_3$, (c-hexyl)SiH$_3$, Ph$_2$SiH$_2$ or PhMeSiH$_2$, these species exhibit no influence on the course of events. Further support for the mechanism shown in Scheme 2 comes when reaction (28) (where M = Hf) is run with two equivalents of PhSiH$_3$. This reaction leads to the formation of phenylsilane dimers and trimers. With time, a new species appears that is probably the tetramer.

Further work required in the area of polysilane self-reaction dehydrocoupling is likely to focus on optimization of catalyst activity, especially with respect to the synthesis of high molecular weight polymers as there is considerable interest in polymers of this type [1]. In addition, the same driving forces for commercialization of high molecular weight polysilanes should also lead to work on the synthesis of linear polygermanes. An additional area of interest is the development of catalysts that will permit the synthesis of tractable polymethylsilane, -[MeSiH]$_x$-, and polysilane itself, -[H$_2$Si]$_x$-. The former is of interest as a preceramic and the latter would have many useful applications in the electronics industry.

Si-H Catalytic Reactions with E-H

The first reports describing catalytic reaction of Si-H bonds with compounds containing acidic hydrogen were those of Chalk [24] and Corriu et al [25]. These reactions involved the alcoholysis of silanes:

$$ROH + R'_3Si\text{-}H \xrightarrow{\quad catalyst \quad} H_2 + R'_3Si\text{-}OR \tag{29}$$

Although these reactions are not useful for the synthesis of organometallic polymers, except perhaps for polysiloxanes (where ROH = H$_2$O and R'$_3$SiH = R'$_2$SiH$_2$), they set the stage for later work wherein the active hydrogen species is an amine [26-30]:

$$RNH_2 + R'_3Si\text{-}H \xrightarrow{\quad catalyst \quad} H_2 + R'_3Si\text{-}NHR \tag{29}$$

The synthesis of polysilazane oligomers via dehydrocoupling has been studied extensively by Laine et al [29, 31-35], using a variety of amine and silane reactants. For example, if Et$_2$SiH$_2$ is reacted with NH$_3$, reaction (30),

$$Et_2SiH_2 + NH_3 \xrightarrow{\quad Ru_3(CO)_{12}/60°C \quad} H_2 + \text{-}[Et_2SiNH]_y\text{-} + H\text{-}[Et_2SiNH]_x\text{-}H \tag{30}$$
$$y = 3\text{-}5 \qquad\qquad x = 3\text{-}5$$

then the major products are primarily cyclomers with the cyclotrisilazane predominating and small quantities of low molecular weight linear oligomers. Typically the catalyst concentration is 0.1 to 0.01 mole percent. Catalyst concentrations at the ppm level can be used for some reactions.

It has not been possible to prepare high molecular weight polysilazanes using reactions analogous to (30) for a variety of reasons. Modeling studies were run using reaction (31), in an effort to find the best conditions for

$$Et_3SiH + RNH_2 \xrightarrow{\quad Ru_3(CO)_{12}/THF/70°C \quad} H_2 + Et_3SiNHR \tag{31}$$
$$R = n\text{-}Pr, n\text{-}Bu, s\text{-}Bu \text{ or } t\text{-}Bu$$

polymer formation [31].

The kinetics and the catalytic cycle(s) for this reaction are extremely complex. In the absence of amine, the silane reacts with $Ru_3(CO)_{12}$ to produce $(Et_3Si)_2Ru_2(CO)_8$, reaction (32), which can be isolated and used in place of $Ru_3(CO)_{12}$. Catalyst concentration studies demonstrate that the rate of

$$6Et_3SiH + 2Ru_3(CO)_{12} \xrightarrow{\;110°C/10\ min\;} 3H_2 + 3(Et_3Si)_2Ru_2(CO)_8 \tag{32}$$

reaction (31) is nonlinearly and inversely dependent on either $[Ru_3(CO)_{12}]$ or $[(Et_3Si)_2Ru_2(CO)_8]$. On a molar basis, $(Et_3Si)_2Ru_2(CO)_8$ is the more active of the two catalyst precursors. These results suggest that the true catalyst forms by cluster fragmentation. Indeed, some evidence was found for the formation of trans-$(Et_3Si)_2Ru(CO)_4$. Thus, higher catalyst concentrations do not improve reaction rates significantly.

A second problem, which also plagues self-reaction dehydrocoupling, is that the reaction rate and product selectivities are extremely susceptible to the steric environment about both silicon and nitrogen. In rate vs $[RNH_2]$ studies, the steric bulk of R controls both the reaction rate and the mechanism. The simple primary amines, $n-PrNH_2$ and $n-BuNH_2$ show an inverse, nonlinear rate dependence on $[RNH_2]$ despite the fact that they are reactants.

By comparison, the $[s-BuNH_2]$ studies reveal a nonlinear positive rate dependence. On moving to the most bulky amine, $t-BuNH_2$, the rate shows almost no dependence on either $[t-BuNH_2]$ or $[Et_3SiH]$. The relative global rates of reaction are $n-PrNH_2 \geq n-BuNH_2 > s-BuNH_2 \gg t-BuNH_2$. The potential product, $(Et_3Si)_2NR$, corroborated by Kinsley et al [30], is never observed with the ruthenium catalyst.

Apparently, the simple primary amines react with $Ru_3(CO)_{12}$ to form fairly stable complexes that deplete the reaction mixture of active catalyst:

$$Ru_3(CO)_{12} + EtCH_2NH_2 \longleftrightarrow (\mu^2\text{-}EtCH{=}NH)H_2Ru_3(CO)_{10} \tag{33}$$

Hence, the inverse dependence on $[n-RNH_2]$. With $s-BuNH_2$, the stability of the amine ruthenium complex is sufficiently reduced such that it plays only a minor role in the catalytic cycle.

With $t-BuNH_2$, there is no α-hydrogen and the amine-ruthenium complex cannot form. However, the fact that the rate is simultaneously independent of both $[Et_3SiH]$ and $[t-BuNH_2]$, is more difficult to explain and suggests that an alternate mechanism is operative. One can envision a situation in which the rate determining step is catalyst activation; however, silyl-ruthenium complexes form readily in the absence of amine. Therefore reaction of the amine at the silicon must be the rate determining step. Crabtree et al have recently studied iridium catalyzed alcoholysis of Et_3SiH and proposed that the reaction proceeds via nucleophilic attack of alcohol on a two-electron three-center Si-H···M intermediate [36]. Perhaps such an intermediate occurs in this instance. Alternately, the slow step may arise as a consequence of cluster fragmentation. The cluster fragmentation possiblity seems reasonable given that the rate vs $[Ru_3(CO)_{12}]$ studies with t-$BuNH_2$ reveal first order dependence on catalyst concentration rather than inverse dependence. These combined results suggest a catalytic cycle such as shown in Scheme 3:

$$M_3 + RCH_2NH_2 \longleftrightarrow H_2(RCH=NH)M_3$$

$$M_3 + 3Et_3SiH \longrightarrow 3HMSiEt_3$$

$$Et_3SiMH + Et_3SiH \longleftrightarrow H_2 + Et_3SiMSiEt_3$$

$$Et_3SiMH + RCH_2NH_2 \longrightarrow Et_3SiNHCH_2R + MH_2$$

and/or

$$Et_3SiMSiEt_3 + RCH_2NH_2 \longrightarrow Et_3SiNHCH_2R + Et_3SiMH$$

$$MH_2 \longleftrightarrow M + H_2$$

Scheme 3.

The severe steric effects observed in reaction (31) are also observed when the dehydrocoupling reaction is used to synthesize oligosilazanes. Reaction (30) exemplifies these effects. This reaction was an attempt to prepare linear, high molecular weight diethylpolysilazanes. Yet, the only products are mixtures of the cyclotrimer and cyclotetramer with low molecular weight linear species ($M_n \approx 500$ D). In contrast, the use of monosubstituted silane precursors provides access to true oligosilazanes as illustrated by reactions (34) and (35):

$$PhSiH_3 + NH_3 \xrightarrow{\text{Ru}_3\text{(CO)}_{12}/60°C} H_2 + \text{-}[PhSiHNH]_x\text{-} \tag{34}$$
$$M_n = 800\text{-}1000 \text{ D}$$

$$n\text{-}C_6H_{13}SiH_3 + NH_3 \xrightarrow{Ru_3(CO)_{12}/60°C} H_2 + \text{-}[n\text{-}C_6H_{13}SiHNH]_x\text{-} \qquad (35)$$
$$M_n \approx 2700 \text{ D}$$

At 60°C, both oligosilazanes are essentially linear. No evidence is found for dehydrocoupling at the tertiary Si-H bonds or at the internal N-H bonds. At 90°C, the tertiary Si-H bonds react to give partially crosslinked oligosilazanes as shown for the phenylpolysilazane in reaction (36):

$$H\text{-}[PhSiHNH]_x\text{-}H + NH_3 \xrightarrow{Ru_3(CO)_{12}/90°C} H_2 + \qquad NH_{0.5}$$
$$|$$
$$\text{-}[PhSiHNH]_x[PhSiNH]_y\text{-} \quad \text{solid, } M_n = 1400 \qquad (36)$$

Reactions (34) and (35) can be run at room temperature, if $Ru_3(CO)_{12}$ is heated (activated Ru Cat) in neat silane prior to addition of NH_3 [35].

In contrast to $PhSiH_3$ and n-hexylSiH_3, $EtSiH_3$ reacts indiscriminantly to give a crosslinked polysilazane that is sufficiently intractable to permit effective characterization. The reactivity differences found for $PhSiH_3$ and n-hexylSiH_3, when compared to $EtSiH_3$, are quite reminiscent of the reactivity differences found for $PhSiH_3$ and n-hexylSiH_3 versus $MeSiH_3$ for self-reaction dehydrocoupling as discussed by Harrod et al [19,20].

The same dehydrocoupling reaction used to synthesize simple oligosilazanes can also be used to further polymerize oligosilazanes produced via other chemical routes. Recent interest in the use of polysilazanes as silicon nitride preceramic polymers [4] originally prompted exploration of dehydrocoupling as a synthetic technique [31-35].

Oligomers of most polysilazanes ceramic precursors have been synthesized by ammonolysis of H_2SiCl_2 or $MeSiHCl_2$, as in reaction (37):

$$H_2SiCl_2 + 3MeNH_3 \xrightarrow{0°C/Et_2O} HNMe\text{-}[H_2SiNMe]_x\text{-}H + 2MeNH_3Cl \qquad (37)$$

Unfortunately, these reactions lead to oligosilazanes with molecular weights too low to have useful polymeric properties. Fortunately, because they contain both Si-H bonds and N-H bonds, ruthenium catalyzed dehydrocoupling can be used to further polymerize these simple olgomers:

$$HNMe\text{-}[H_2SiNMe]_x\text{-}H \xrightarrow{Ru_3(CO)_{12}/60\text{-}90°C} \text{polymers} \qquad (38)$$

When reaction (38) (where $x \approx 20$, $M_n \approx 1200$ D) is run over a 65 h period, M_n increases to only 2300 D with $M_W \approx 25,000$ D. The polysilazane viscosity changes from approximately 1-3 poise to 100 poise in the same time period. These results are typical of a gelation process. However, the GPC curve for the 65 h run [37], indicates that a significant portion of the polymer has a molecular weight well above 10K daltons.

The GPC curve is bimodal which suggests more than one mechanism for polymerization. In recent work, Youngdahl et al [35] find evidence for an additional mechanism for oligosilazane coupling. They find that the amine capped oligosilazanes produced in reaction (37), normally stable to >100°C, will condense in the presence of catalyst, reaction (39):

$$2 \text{ --SiNHMe} \xrightarrow{\text{activated Ru Cat/40°C}} \text{MeNH}_2 + \text{--Si-N(Me)-Si--} \qquad (38)$$

Two other catalytic methods of synthesizing oligo- and polysilazanes have also been discovered as discussed in the following sections.

Redistribution Reactions

Redistribution reactions, as defined here, involve the exchange of ligands or moieties between silicon centers, reaction (39):

$$\text{SiX}_n + \text{SiY}_m \xleftarrow{\quad\text{catalyst}\quad} \text{SiX}_{n-z}\text{Y}_z + \text{SiY}_{m-z}\text{X}_z \qquad (39)$$

The use of n and m in place of 4 allows for the possibility that such catalytic exchanges may occur between penta- or hexacoordinated silicon species in addition to four coordinate silicon compounds.

Transition-metal promoted catalytic redistribution on silicon has already been well reviewed by Curtis and Epstein [15]. Some overlap between the two reviews is necessary given the different objectives; however, the majority of the work discussed below is new. In particular, the interest in this section is on the redistribution reactions of Si-H and Si-O bonds to form silisesquioxane polymers.

Curtis et al [15] have examined the utility of group 8 metal catalyzed redistribution of hydridosiloxanes, tetramethyldisiloxane in particular:

$$\text{HMe}_2\text{Si-O-SiMe}_2\text{H} \xrightarrow{\quad\text{(Ph}_3\text{P)}_2\text{Ir(CO)Cl/60°C}\quad}$$
$$\text{Me}_2\text{SiH}_2 + \text{H(Me}_2\text{SiO)}_n\text{-SiMe}_2\text{H} \;(n = 1\text{-}6) \qquad (40)$$

In addition to Me_2SiH_2 and simple oligomers, the iridium catalyst used in reaction (40) also promotes methyl group transfer to form Me_3Si- capped oligomers. Although higher oligomers are also produced, their concentrations are small. The mechanism has already been discussed in detail and will not be covered here except to note that Curtis and Epstein state that there are discrepancies between what is suggested and the experimental evidence, especially with regard to alkyl transfers. The problem of a four electron redox process, as mentioned above, is also discussed. Further work on reaction (40) is clearly of interest in light of these discrepancies, Luo and Crabtree's results [36], and the possibility of binuclear reactions, as discussed above.

As with group 8 metal catalyzed self-reaction dehydrocoupling, group 8 metal catalyzed redistribution of hydridosiloxanes gives only oligomeric species. If the early transition metals, in particular Harrod's catalyst are used, considerable changes in product distribution are obtained.

Harrod et al [38] first described titanium catalyzed redistribution of simple hydridosiloxanes, e.g. $(EtO)_2MeSiH$ in 1986 and have recently followed up with a more detailed communication on the subject [39]:

$$(EtO)_2MeSiH \xrightarrow{\quad Cp_2TiMe_2/RT \quad} (EtO)_3SiMe + MeSiH_3 \qquad (41)$$

This reaction, which can be run neat or in a variety of solvents with less than 0.01 mole percent catalyst, can be used very successfully to form polydimethylsiloxane polymers from $HMe_2SiOSiMe_2H$, as in reaction (40), with Cp_2TiMe_2 catalyst. The polymers have $M_n \approx 10K$ D and a polydispersity of 1.6. The proton and silicon NMRs are indistinguishable from commercially produced polydimethylsiloxane.

Methysilsesquioxane polymers can also be produced in a similar manner if the reactants contain -[MeHSiO]- groups, reaction (42). The source of the -[MeHSiO]- groups can either be cyclomers or linear oligomers [40]. The

$$-[MeHSiO]_x- \xrightarrow{\quad Cp_2TiMe_2/RT/N_2 \quad} MeSiH_3 + -[MeSi(O_{1.5})]_x- \qquad (42)$$

titanocene catalyst is heat and light sensitive and, somewhat air sensitive. Reactions such as (42) typically have an induction period which can range from minutes to hours depending on the amount of oxygen present in the system. This induction period appears to be solely for catalyst activation. Once the reaction has started, the initial yellow color of the dimethyl titanocene

changes rapidly to royal blue and the reaction is complete in a matter of
minutes. No matter what the starting compound, if the reaction is run neat, the
final product has a composition that is about $-[MeHSiO]_{0.35}[MeSi(O_{1.5})]_{0.65}-$.
This polymer, when heated to $>200°C$, loses all of its $-[MeHSiO]-$ groups to form
a polymer consisting entirely of $-[MeSi(O_{1.5})]_x-$, methylsilsesquioxane. The
$-[MeSi(O_{1.5})]_x-$ polymer is stable to greater than $500°C$ in nitrogen [40].

Because of the interest in new routes to polysilazanes, attempts were
made to use the Harrod catalyst for redistribution of the silazane, $-[MeHSiNH]_x-$
[41]. Unfortunately, the titanocene catalyst is inactive for redistribution when
oxygen is replaced by nitrogen; although the solutions do turn blue. What is
intriguing about this system is that if dimethyl titanocene is added to mixtures
of $-[MeHSiO]-$ and $-[MeHSiNH]_x-$, then polymerization at room temperature does
occur. In fact, the MeHSiO : MeHSiNH ratio can be varied from 1:1 to 1:50
without affecting the catalyst's activity. This catalyst system provides a new
route to polysilazane polymers.

Harrod et al [39] proposed two mechanisms for reaction (40), Scheme 4

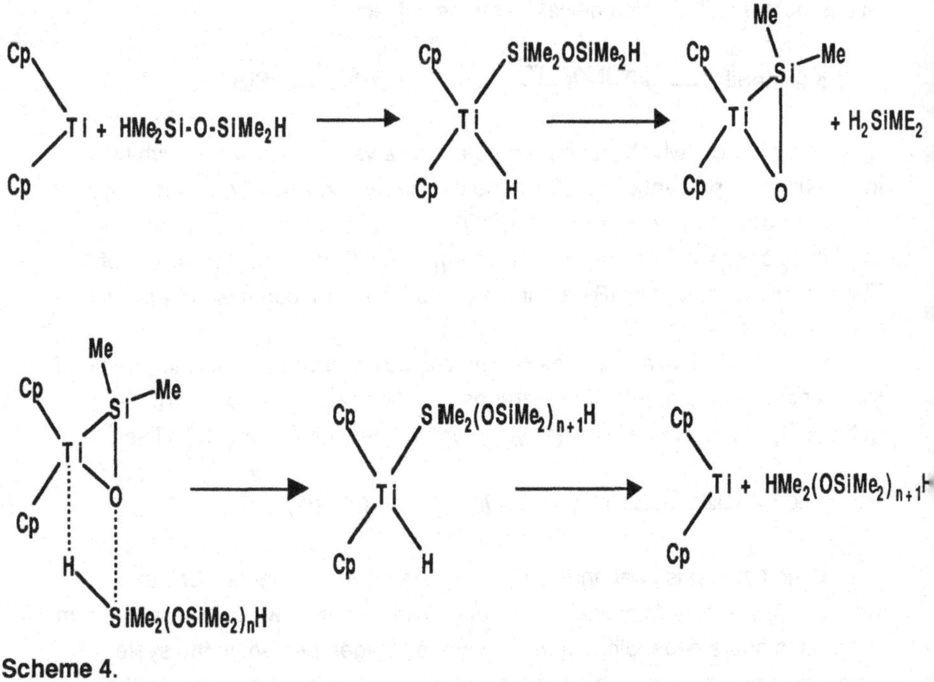

Scheme 4.
and Scheme 5, that may explain what occurs during redistribution catalysis. In

Scheme 4 the active catalyst is a mononuclear, Ti (IV) complex. In Scheme 5 catalysis requires two Ti (III) metal centers. The authors prefer the Scheme 5 mechanism because all of the species isolated from solution to date (see above) are Ti (III) species:

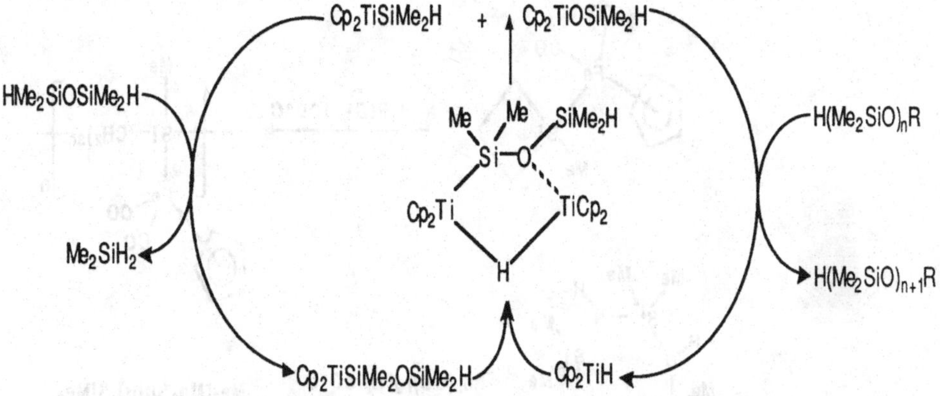

Scheme 5.

Considerably more work needs to be done on these reactions, especially in view of the silazane results, before a detailed understanding of the catalytic cycle is possible. These findings are of significant commercial interest, because the titanium catalysts offer much higher activity than possible with the group 8 metal catalysts as well as good product selectivity and, entrée to new, potentially useful silsesquioxane polymers.

Ring-Opening Catalysis

Transition metal catalyzed ring opening is the oldest reaction known for the catalytic synthesis of organometallic polymers [42,43]. Again, the only known examples involve heterocyclic monomers that contain silicon, cyclic carbosilanes or silazanes. Ring-opening polymerization of cyclic carbosilanes was discussed by Curtis and Epstein [15]. As such, discussion here is limited primarily to new developments.

Reactions (43)-(45) are examples of transition metal catalyzed ring opening polymerization [34,42-47]. Reaction (43), in the absence of a silane, provides gummy high polymers with molecular weights of 10^5 to 10^6 D. Addition of silane permits the synthesis of lower molecular weight silane capped polymers and oligomers. Reaction (44) is exceedingly interesting

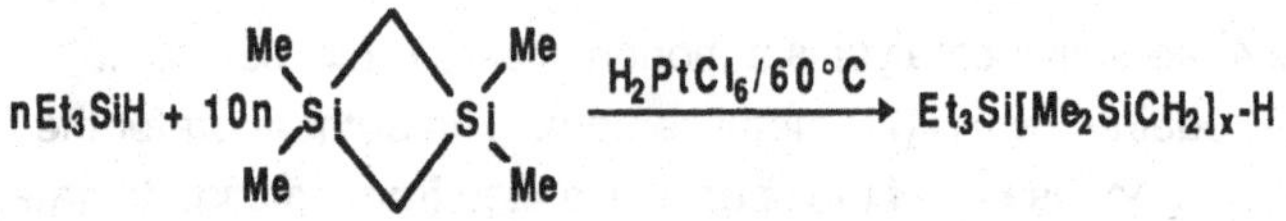

$$n Et_3SiH + 10n \; \underset{Me}{\overset{Me}{Si}} \underset{Me}{\overset{Me}{Si}} \xrightarrow{H_2PtCl_6/60°C} Et_3Si[Me_2SiCH_2]_x\text{-}H \qquad (43)$$

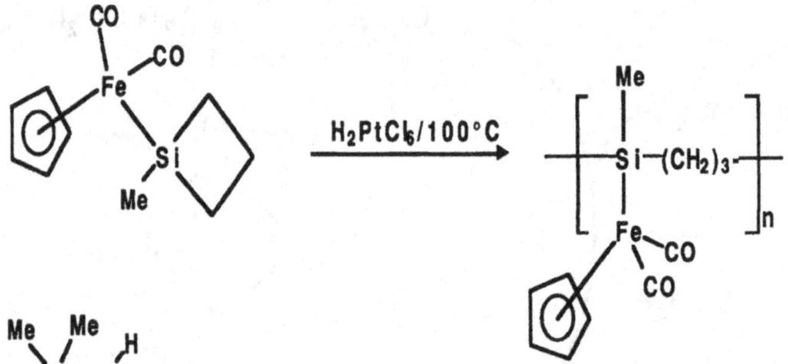

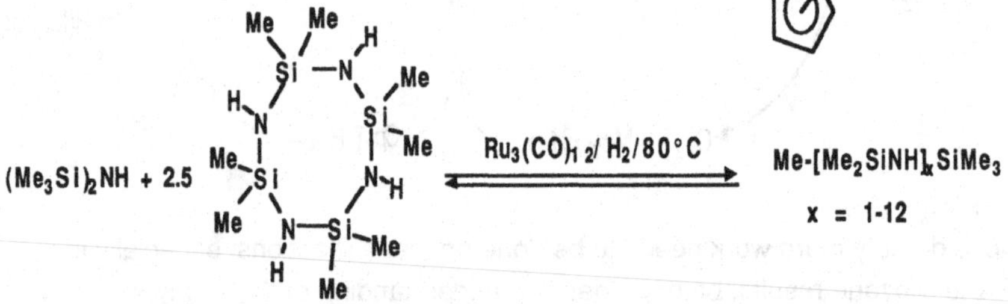

$$\xrightarrow{H_2PtCl_6/100°C} \left[\underset{\underset{Fe-CO}{\overset{CO}{|}}}{\overset{Me}{\underset{|}{Si}}}\text{-}(CH_2)_3 \right]_n \qquad (44)$$

$$(Me_3Si)_2NH + 2.5 \quad \underset{\text{ring structure}}{} \quad \underset{H_2/80°C}{\overset{Ru_3(CO)_{12}}{\rightleftharpoons}} \quad Me\text{-}[Me_2SiNH]_x SiMe_3 \qquad (45)$$

$$x = 1\text{-}12$$

because it leads to one of the first organometallic polymers containing a pendant organometallic ligand. Reaction (44) gives similar materials although no molecular weights are reported [46]. Reaction (45) can also be run in the absence of capping agent, $(Me_3Si)_2NH$. This leads to higher molecular weight, hydrogen capped oligomers, albeit in much lower yields.

Reactions (43) and (44) are normally run at 50-150°C with less than 1 mole percent catalyst. Most of the group 8 metals will catalyze these reactions; however, platinum compounds afford the best catalyst activities.[44]

Although, several mechanisms have been proposed for reactions (43) and (44) [15,44.45]; a mechanism that fits all the facts has not been forthcoming. For example, Cundy et al [45] have suggested a catalytic sequence involving chlorine transfer polymerization; however, many catalysts and cyclic carbosilanes undergo ring opening polymerization in the absence of chlorine. Thus, such a mechanism is not likely to serve as a general model.

Scheme 6 presents a reasonable mechanistic explanation for what has been reported to date; however, it should not be considered definitive:

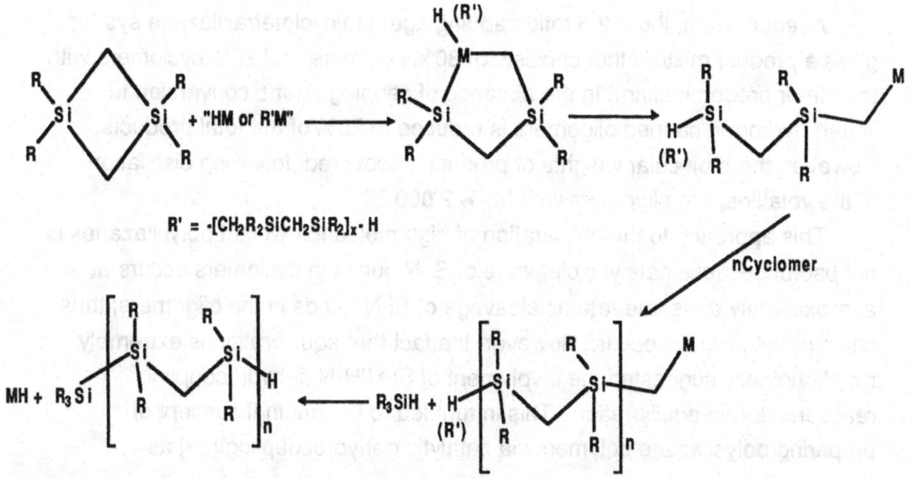

$$R' = -[CH_2R_2SiCH_2SiR_2]_x- H$$

Scheme 6.

The mechanism for catalytic ring opening oligomerization of cyclo-silazanes, reaction (45), as shown in Scheme 7, appears to be similar to that suggested in Scheme 6, for cyclic carbosilanes. In both cases, it is likely

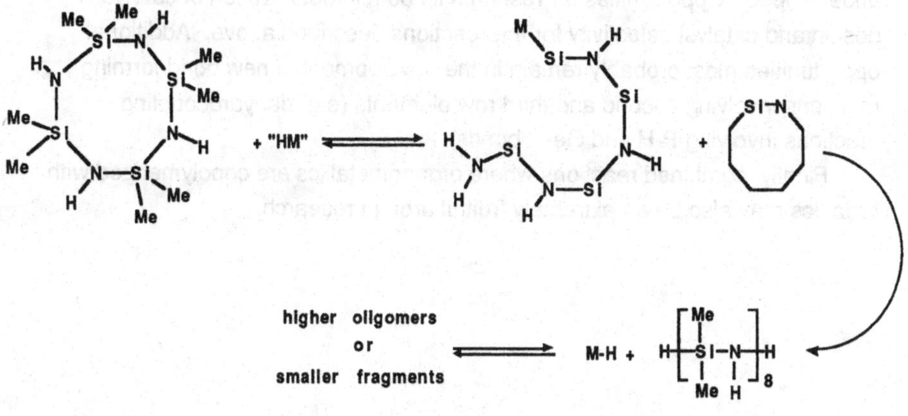

Scheme 7.

that metal hydride are necessary interemediates for effective polymerization.

In fact, the introduction of 1 atm of H_2 to reaction (45), permits equilibration in 1 h at 80°C; whereas, without H_2, equilibrium is not obtained even after 24 h at 135°C. Furthermore, the $Ru_3(CO)_{12}/H_2$ (1 atm) catalyst system can be replaced with $H_4Ru_4(CO)_{12}$ without affecting the reaction rate.

At equilibrium, the 1:2.5 ratio capping agent to cyclotetrasilazane system gives a product mixture that consists of 80% polymers and 20% cyclomers with the trimer predominating. In the absence of capping agent, conversion to linear, hydrogen capped oligomers is reduced to 20% of the total products. However, the molecular weights of products recovered, following distillation of the volatiles, are oligomers with $M_n \approx 2,000$ D.

This approach to the preparation of high molecular weight polysilazanes is not useful because catalytic cleavage of Si-N bonds in cyclomers occurs at approximately the same rate as cleavage of Si-N bonds in the oligomers; thus, chain growth will not occur. However, the fact that equilibration is extremely rapid originally suggested the involvment of Si-H/H-N dehydrocoupling reactions during equilibration. This in turn led to the original concept of preparing polysilazane polymers via catalytic dehydrocoupling [29] as discussed above.

Future Directions

The potential utility of catalysis as a synthetic tool for the preparation of organometallic polymers is just being realized. Much of this potential remains undeveloped. Opportunities for research lie both in optimization of catalyst design and catalyst selectivity for the reactions described above. Additional opportunities most probably remain in the development of new bond forming reactions involving second and third row elements (e.g. dehydrocoupling reactions involving P-H and Ge-H bonds).

Finally, combined reactions where organometallics are copolymerized with organics may also be an extremely fruitful area of research.

Acknowledgments
We would like to thank many colleagues for the preprints, reprints and helpful discussions that served as the basis of this review. We especially wish to thank Professors Harrod, Tilley and Sneddon for timely discussions. We are also grateful to the Strategic Defense Sciences Office through the Office of Naval Research Contract for support of the work presented here.

References

1. a. R. D. Miller, J. F. Rabolt, R. Sooriyakumaran, W. Fleming, G. N. Ficke, B. L. Farmer, and H. Kuzmany in "Inorganic and Organometallic Polymers" Am. Chem. Soc. Symp. Ser. **360**, ed. M. Zeldin, K. Wynne, and H. Allcock, 1988 p 43. b. R. D. Miller, G. Wallraff, N. Clecak, R. Sooriyakumaran, J. Michl, T. Karatsu, A. J. McKinley, K. A. Klingensmith, J. Downing, Polym. Mater. Sci. Eng., **60**, 49 (1989).

2. a. H. R. Allcock in "Inorganic and Organometallic Polymers" Am. Chem. Soc. Symp. Ser. **360**, ed. M. Zeldin, K. Wynne, and H. Allcock, 1988 p 251. b. S. Ganapathiappan, K. Chen, and D. F. Shriver, J. Am. Chem. Soc., **111**, 4091 (1989).

3. a. "Nonlinear Optical Effects in Organic Polymers" ed. J. Messier, F. Kajzar, P. Prasad, D. R. Ulrich, NATO ASI Ser.; Series E: Appl. Sci. **162**, Kluwer Publ.1989. b. D. R. Ulrich, Mol. Cryst. Liq. Cryst., **160**, 1 (1988).

4. See chapters 10-12 and 30-32 in "Inorganic and Organometallic Polymers" Am. Chem. Soc. Symp. Ser. **360**, ed. M. Zeldin, K. Wynne, and H. Allcock,1988.

5. See Part VII of "Better Ceramics Through Chemistry III" Mat. Res. Soc. Symp. Proc.**121**, ed. C. J. Brinker, D. E. Clark, D. R. Ulrich, 1988.

6. E. W. Corcoran, Jr. and L. G. Sneddon, J. Am. Chem. Soc., **106**, 7793 (1984).

7. Y. D. Blum and R. M. Laine, U. S. Patent 4,801,439 Jan, 1989.

8. A. T. Lynch and L. G. Sneddon, Inorg. Chem. Paper No. 296, paper presented at the American Chemical Society, Fall Meeting, Los Angeles, CA 1988

9. L. G. Sneddon private communication.

10. R. Wilczynski and L. G. Sneddon, Inorg. Chem., **20**, 3955 (1981).

11. I. Ojima, S. I. Inaba, T. Kogure, and Y. Nagai, J. Organomet. Chem., **55**, C7 (1973).

12. a. J. Y. Corey, L. S. Chang, E. R. Corey, Organomet., **6**, 1595 (1987). b. L. S. Chang and J. Y. Corey,, Organomet., **8**, 1885 (1989).

13. K. A. Brown-Wensley, Organomet., **6**, 1590 (1987).

14. H.-G. Woo and T. D. Tilley in, "Fourth International Conference on

Ultrastructure Processing of Polymers, Ceramics and Glasses" Tucson, Az; February, 1989, eds. D. Uhlmann and D. R. Ulrich, symp. proc. in press.

15. M. D. Curtis and P. S. Epstein, Adv. Organomet. Chem., **19**, 213 (1981).

16. C. Zybill, D. L. Wilkinson, C. Leis and G. Müller, Angew. Chem., **101**, 206 (1989).

17. a. C. Aitken, J. F. Harrod, E. Samuel, J. Organomet. Chem., **279**, C11 (1985). b. C. Aitken, J. F. Harrod, E. Samuel, J. Am. Chem. Soc., **108**, 4059 (1986). c. J. F. Harrod and S. S. Yun, Organomet., **6**, 1381 (1987). d. C. Aitken, J. F. Harrod, U. S. Gill, Can. J. Chem., **65,** 1804 (1987).

18. a. J. F. Harrod in "Inorganic and Organometallic Polymers", A. C. S. Symposium Series **360**, 89 (1988). b. C. Aitken, J.-B. Barry, F. Gauvin, J. F. Harrod, A. Malek, and D. Rosseau, Organomet., **8**, 1732 (1989).

19. J. F. Harrod in "Transformation of Organometallics into Common and Exotic Materials: Design and Activation," NATO ASI Ser. E: Appl. Sci.-No. **141**, R. M. Laine Ed.; Martinus Nijhoff Publ., Amsterdam (1988) p 103.

20. C. Aitken, J. F. Harrod, A. Malek and E. Samuel, J. Organomet. Chem., **349**, 285 (1978).

21. A. Marchand, P. Gerval, P. Rivière and J. Satgé, J. Organomet. Chem., **162**, 365 (1978).

22. H.-G. Woo and T. D. Tilley, J. Am. Chem. Soc., **111**, 3757 (1989).

23. See for example P. L. Watson, J. Am. Chem. Soc., **105**, 6491 (1983).

24. A. J. Chalk, J. Chem. Soc. Chem. Commun., 847 (1967).

25. R. J. P. Corriu and J. J. E. Moreau, J. Organomet. Chem., **114**, 135 (1967).

26. R. C. Borchert, U. S. Patent 3,530,092 issued Sept.1970

27. L. H. Sommer and J. D. Citron, J. Org. Chem., **32**, 2470 (1979).

28. H. Kono and I. Ojima, Organ. Prep. and Proc., **5**, 135 (1973).

29. Y. D. Blum and R. M. Laine, Organomet., **5**, 2801 (1986).

30. K. K. Kinsley, T. J. Nielson and T. J. Barton, Main Group Met. Chem., **10**, 307 (1987).

31. Y. D. Blum, R. M. Laine, K. B. Schwartz, D. J. Rowcliffe, R. C. Bening and D. B. Cotts, Better Ceramics Through Chemistry II, Mat. Res. Symp. Proc. Vol. **73**, ed. C. J. Brinker, D. E. Clark, and D. R. Ulrich, 1986, pp 389.

32. C. Biran, Y. D. Blum, R. M. Laine, R. Glaser, and D. S. Tse, J. Mol. Cat., **48**, 183 (1989).

33. R. M. Laine, Y. Blum, D. Tse, and R. Glaser, "Inorganic and Organometallic Polymers.", Am. Chem. Soc. Symp. Ser. Vol. **360**, K. Wynne, M. Zeldin and H. Allcock Ed. (1988) pp 124-142.

34. R. M. Laine, Platinum Met. Review, **32**, 64 (1988).
35. K. A. Youngdahl, R. M. Laine, R. A. Kennish, T. R. Cronin, and G. A. Balavoine, in <u>Better Ceramics Through Chemistry III</u>, Mat. Res. Symp. Proc. Vol. **121**, C. J. Brinker, D. E. Clark, and D. R. Ulrich Eds., 489 (1988).
36. X.-L. Luo and R. H. Crabtree, J. Am. Chem. Soc., **111,** 2527 (1989).
37. A. W. Chow, R. D. Hamlin, Y. Blum and R. M. Laine, J. Polym. Sci. Part C: Polym. Lett., **26,** 103-108 (1988).
38. J. F. Harrod, S. Xin, C. Aitken, Y. Mu and E. Samuel, International Conference on Silicon Chemistry, June,1986; St. Louis, Mo.
39. J. F. Harrod, S. Xin, C. Aitken, Y. Mu, and E. Samuel, submitted to Can. J. Chem. (1989).
40. R. M. Laine, K. A. Youngdahl, F. Babonneau, J. F. Harrod, M. Hoppe and J. A. Rahn submitted to J. Chem. Mat., (1989)
41. K.A. Youngdahl, M. L. Hoppe, R. M. Laine, J. A. Rahn and J. F. Harrod; <u>Fourth Internat. Conf. on Ultrastructure Processing of Glasses, Ceramics, Composites and Polymers</u>, eds. D. Uhlmann and D. R. Ulrich, symp. proc. in press.
42. N. S. Nametkin, V. M. Vdovin, and P. L. Grinberg, Izv. Akad. Nauk. SSSR, Ser. Khim., 1133 (1964).
43. D. R. Weyenberg and L. E. Nelson, J. Org. Chem., **30**, 2618 (1965).
44. W. R. Bamford, J. C. Lovie and J. A. C. Watt, J. Chem. Soc. (C) 1137 (1966).
45. C. S. Cundy, C. Eaborn and M. F. Lappert, J. Organomet. Chem. **44**, 291 (1972).
46. C. S. Cundy, M. F. Lappert and C.-K. Yuen, J. Chem. Soc. Dalton, 427 (1978).
47. E. Bacque, J.-P. Pillot, M. Birot and J. Dunoguès, in "Transformation of Organometallics into Common and Exotic Materials: Design and Activation," NATO ASI Ser. E: Appl. Sci.-No. **141**, R. M. Laine Ed.; Martinus Nijhoff Publ., Amsterdam (1988) p 116.

Homogeneous Catalytic Hydrogenation of Aromatic Hydrocarbons and Heteroaromatic Nitrogen Compounds: Synthetic and Mechanistic Aspects.

Richard H. Fish

Lawrence Berkeley Laboratory

University of California

Berkeley, CA 94720

Abstract:

The synthetic and mechanistic aspects of the homogeneous catalytic hydrogenation of mono and polynuclear aromatic hydrocarbons and the corresponding heteroaromatic nitrogen compounds will be reviewed. A comparison of the regioselectivities under various hydrogenation conditions for both classes of compounds will be discussed for a wide variety of transition-metal complexes with regard to substrate binding at the metal center and the role of free radical intermediates in metal carbonyl hydride reactions, where substrate does not bind to the metal center prior to hydrogen transfer. The polynuclear heteroaromatic nitrogen compounds appear to hydrogenate more readily, under similar reaction conditions, than the polynuclear aromatic hydrocarbons. The relative rates of hydrogenation of a variety of heteroaromatic nitrogen compounds as well as compounds that inhibit and enhance selective hydrogenation of the nitrogen-containing ring will be addressed. A relatively new spectroscopic technique, high pressure nuclear magnetic resonance spectroscopy, will be shown to be a powerful tool to elucidate the mechanisms of the regioselective hydrogenation of polynuclear heteroaromatic nitrogen compounds.

R. Ugo (ed.), Aspects of Homogeneous Catalysis, Vol. 7, 65–83.

Introduction

The use of homogeneous catalysts to effect the selective hydrogenation of polynuclear aromatic and heteroaromatic nitrogen compounds found its impetus in studies on the use of these compounds as models for similar structures found in coal. The practical importance being that hydrogen up-grading and removal of nitrogen as well as sulfur from fossil fuels are dependent somewhat on the regioselectivity of the hydrogenation reactions. Since transition-metal, homogeneous hydrogenation catalysts are known to operate at lower temperatures and lower pressures of hydrogen gas in comparison to their heterogeneous counterparts and, more importantly, are more selective, this then makes them interesting complexes to study from a synthetic and mechanistic point of view.[1]

In this chapter, the synthetic scope and known mechanistic aspects of the homogeneous catalytic hydrogenation chemistry of aromatic hydrocarbons and the corresponding heteroaromatic nitrogen compounds will be reviewed. This will include the studies using Fe, Co, Mn, Rh, and Ru complexes under water gas shift (CO, H_2O); synthesis gas (CO, H_2); and hydrogen (H_2, alone) reaction conditions. The bonding mode of the polynuclear aromatic and heteroaromatic nitrogen ligands will be also be addressed to ascertain its important role in the regiochemistry of these selective reductions. The mechanisms of homogeneous catalyzed reductions are often characterized by kinetic evidence and the use of deuterium labelling as well as other spectroscopic information that may not always clearly define the exact pathway of the catalytic reaction. The recent advent of high pressure NMR spectroscopy will provide needed information about homogeneous hydrogenation mechanisms in real-time, and an example of the usefulness of this powerful technique will be presented for the regioselective reduction of quinoline to 1,2,3,4-tetrahydroquinoline with a cationic organorhodium complex.

Hydrogenation of Mono and Polynuclear Aromatic Hydrocarbons

The earlier reports by Friedman and Wender and their co-workers in this relatively unstudied area of homogeneous catalysis were concerned with polynuclear aromatic compounds such as phenathrene with CO and H_2 in the presence of $Co_2(CO)_8$ (eq1).[2]

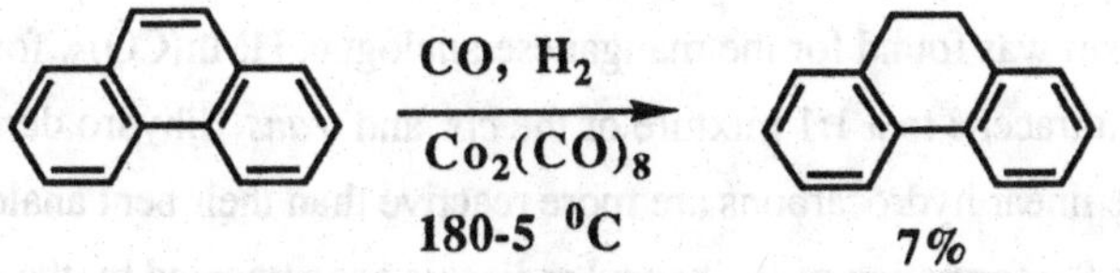

$$\text{phenanthrene} \xrightarrow[\substack{Co_2(CO)_8 \\ 180\text{-}5\ ^0C}]{CO,\ H_2} \text{product} \quad 7\% \tag{1}$$

However, the linear polynuclear aromatic compounds such as anthracene
were found to be more active under the hydrogenation conditions and at lower temperatures
(eq 2).[2]

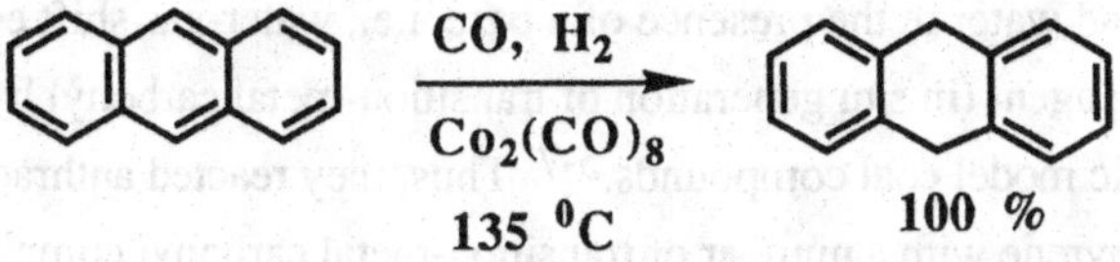

$$\text{anthracene} \xrightarrow[\substack{Co_2(CO)_8 \\ 135\ ^0C}]{CO,\ H_2} \text{product} \quad 100\ \% \tag{2}$$

The results can be rationalized by the mechanism of hydrogenation, which was elucidated
by several groups[3a-c] and entails the formation of $(CO)_4CoH$ followed by hydrogen
radical addition to the aromatic nucleus to give a carbon radical.[3a] The carbon radical then
abstracts a hydrogen from another cobalt hydride to form product. The free radical
mechanism is also supported by reduction of 9,10-dimethylanthracene to provide a ~1:1
mixture of *cis* and *trans*- 9,10-dihydro-9,10-dimethylanthracene (eq3).[3c]

$$H_2 + Co_2(CO)_4 \rightleftharpoons 2HCo(CO)_4$$

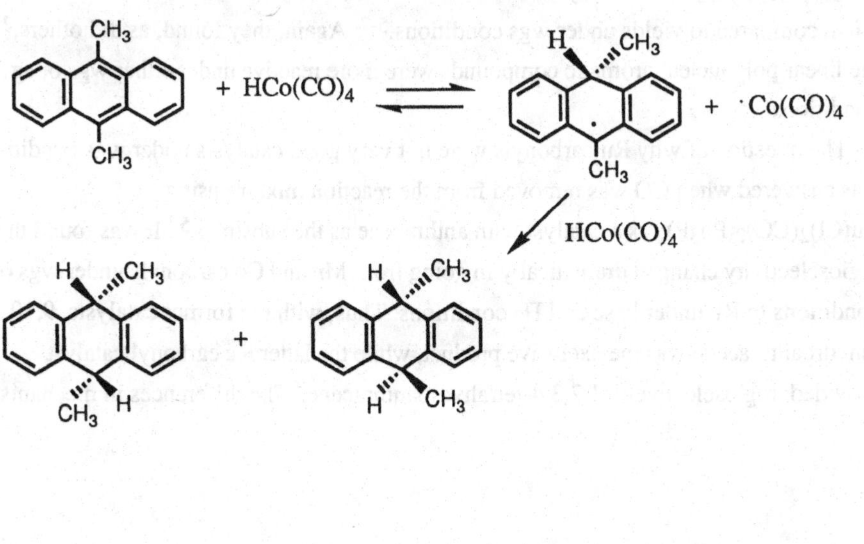

$$\text{9,10-dimethylanthracene} + HCo(CO)_4 \rightleftharpoons \text{radical intermediate} + \cdot Co(CO)_4 \tag{3}$$

$$\downarrow HCo(CO)_4$$

cis and trans products

A similar mechanism was found for the manganese analogue, $HMn(CO)_5$, for the reduction of 9,10-dimethylanthracene to a 1:1 mixture of the *cis* and *trans* dihydro derivatives.[4] One reason that the linear hydrocarbons are more reactive than their bent analogues is that in these latter cases the formation of the benzyl radical is accompanied by the stabilization of two phenyl groups not one (eq 3). This lowers the activation energy for H· radical addition and has a profound effect on the rate of reaction.

Several years ago, Fish and co-workers were intrigued by the possibility of using carbon monoxide and water in the presence of a base, i.e., water-gas shift conditions (wgs), as a reducing agent (in situ generation of transition-metal carbonyl hydrides) for polynuclear aromatic model coal compounds.[5-7] Thus, they reacted anthracene, phenanthrene, and pyrene with a number of transition-metal carbonyl complexes $[M_X(CO)_Y$ M = Fe, Co, Mn, Rh, Ru, W, Mo, Cr,] and found that only anthracene was reduced to 9,10-dihydroanthracene with Mn, Co, and Fe carbonyls, while the other polynuclear aromatic compounds and transition-metal carbonyls studied were inactive. They concluded that this metal carbonyl reactivity was based on the least reactive water-gas shift catalysts having the best hydrogenation activity, i.e., hydrogenation of polynuclear aromatic substrate being competitive with loss of hydrogen gas, and also provided clear evidence that the hydrogen came from H_2O by substituting D_2O for H_2O and finding 9,10-dideuteroanthracene.[5]

Fish and co-workers also found that under synthesis gas (sg) conditions (CO, H_2), with $Mn_2(CO_8(Bu_3P)_2$ as the catalyst, better yields of 9,10-dihydroanthracene were obtained when compared to yields under wgs conditions.[5-7] Again, they found, as did others,[2] that the linear polynuclear aromatic compounds were more reactive under either wgs or sg conditions.

The question of why Ru carbonyls were not very good catalysts under wgs conditions was answered when CO was removed from the reaction mixture using $Ru(Cl)_2(CO)_2(Ph_3P)_2$ as a catalyst with anthracene as the substrate.[5] It was found that the regioselectivity changed dramatically in going from Mn and Co carbonyls under wgs or sg conditions to Ru under base and H_2 conditions. Thus, with the former catalysts, 9,10-dihydroanthracene was the exclusive product, while the latter Ru carbonyl catalyst provided, regioselectively, 1,2,3,4-tetrahydroanthracene. The differences in mechanism were evident in that with the Ru carbonyls substrate binding to the metal center was a prerequisite for reduction to occur, while with the Mn and Co carbonyls this was not necessary, since changes in the carbon monoxide pressures were independent of product formation.

Consequently, carbon monoxide must act as a competitive inhibitor with polynuclear aromatic substrates for the Ru metal center. Halpern and co-workers have studied the mechanism of selective anthracene hydrogenation with Ru polyhydrides. They were able to clearly show that anthracene binds η^4 to the Ru metal center, via the reaction of fac-$[RuH_3(PPh_3)_3]^-$ with anthracene, followed by the isolation and identification of $[RuH(Ph_3P)_2$anthracene]K, and demonstrate that it forms 1,2,3,4-tetrahydroanthracene upon reaction with H_2 (eq 4).[8a] Halpern and Landis have also determined that the $[Rh(diphos)(\eta^6$-anthracene)]^+$ complex reacts with H_2 to give 1,2,3,4-tetrahydroanthracene, again via a presumed η^4 hydride intermediate.[8b]

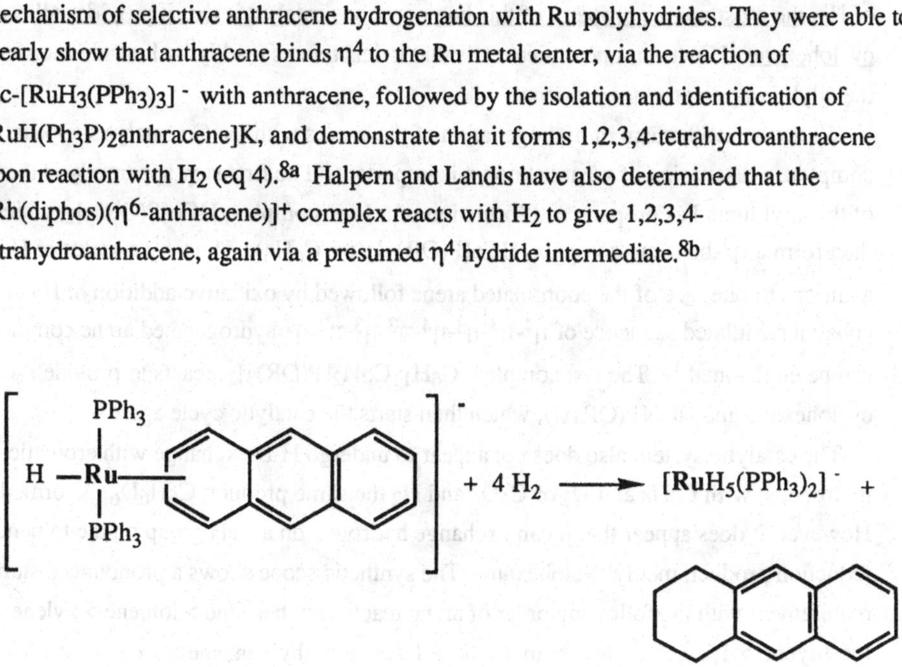

$$\left[\begin{array}{c} PPh_3 \\ | \\ H - Ru \cdots \\ | \\ PPh_3 \end{array} \right]^- + 4\,H_2 \longrightarrow [RuH_5(PPh_3)_2]^- + \qquad (4)$$

The more difficult to hydrogenate mononuclear aromatic compounds, such as benzene, have been the object of several studies and a brief review of this area would be of interest. While benzene and its derivatives were readily hydrogenated to cyclohexane with heterogeneous catalysts, similar reactivity with homogeneous catalysts has proven to be more difficult. One reason is that providing definitive evidence for a homogeneous reaction that is free of any "heterogeneous component" is not always easy, and another is the high activation energy barrier for arene ring hydrogenation that is needed in order to overcome the resonance stabilization energies of arenes.

Among the more important studies in the area of benzene hydrogenation, were those of Muetterties and co-workers.[9,10] They discovered that simple organocobalt compounds of general formula, η^3-C$_3$H$_5$Co[P(OR)$_3$]$_3$ (R= CH$_3$, C$_2$H$_5$, C$_3$H$_7$), readily hydrogenated aromatic hydrocarbons in a steroselective manner at ambient temperature and low pressures of H$_2$; the most unique property of this homogeneous catalyst is that it provides all *cis* cyclohexane. However, one drawback is that it deactivates readily by hydride transfer to the allyl group to produce propene.

While the mechanism of hydrogenation of the arenes with the Co catalysts appears complex, a discussion of what is known is important. The first step apparently is the loss of the allyl ligand as propane and formation of a 14 e$^-$ complex, HCo[P(OR)$_3$]$_2$, which can then form a η^4-benzene complex, HCo[P(OR)$_3$]$_2$(η^4-C$_6$H$_6$). Thus, by a series of H additions to one face of the coordinated arene followed by oxidative addition of H$_2$ to cobalt a postulated sequence of η^4-η^3-η^3-η^4-η^3-η^3-η^2-η^1 hydrogenated arene complexes can be envisioned.[10] The last complex, C$_6$H$_{11}$CoH$_2$[P(OR)$_3$]$_2$, reacts to provide *cis* cyclohexane and HCo[P(OR)$_3$]$_2$, which then starts the catalytic cycle again.

The catalytic system also does not appear to undergo H-D exchange with aromatic hydrogens; with C$_6$H$_6$ and D$_2$ or C$_6$D$_6$ and H$_2$ the same product, C$_6$H$_6$D$_6$, is formed. However, it does appear that it can exchange hydrogen on a CH$_3$ group on the toluene reduction product, methylcyclohexane. The synthetic scope shows a pronounced steric requirement with the following order of arene reactivity: benzene > toluene > xylene > mestilyene > 1,2,4,5-tetramethylbenzene > 1,2,3-trimethylbenzene >>> hexamethylbenzene. The electronic effect is also pronounced with electron-withdrawing groups, F, CN, and NO$_2$, on an arene ring causing the systems to be unreactive.

In a review article by Jonas that is related to the above-mentioned Muetterties results, he points out some interesting unpublished thesis studies from his Laboratory on the use of the cobalt complex, (η^3-cyclooctenyl)CoP$_2$(CH$_2$)$_3$, which acts as a catalyst precursor for benzene hydrogenation.[10a] Jonas claims that the mechanism includes the hydrogenolysis of the cyclooctenyl ligand to provide a polynuclear cobalt hydride that reacts with benzene, via a hydrocobaltation, to give a (η^5-cyclohexadienyl)cobalt complex. This further reacts with H$_2$ gas to give a (η^4-1,3-cyclohexadienyl)cobalt hydride, which then reacts with more

H_2 gas to provide cyclohexane. If these results can be verified, they could represent one of the few mechanistically viable pathways for benzene hydrogenation.

Among the many factors that are important for arene hydrogenation to proceed is η^4 coordination, which lowers the aromatic resonance stabilization energy (bent arene) and provides a driving force for arene ring hydrogenation. The formation of η^4 as well as η^2 arene complexes may well be critical for any successful arene hydrogenation catalyst. For example, bis(hexamethylbenzene)ruthenium(0), with one arene ring in a η^4 coordination and the other η^6 is a long-lived catalyst for arene ring hydrogenation; it is thought that η^4 coordination (16e⁻) is important in that it can oxidatively add H_2.[11,12] In related studies, Bennett and co-workers discovered that a (hexamethylbenene)ruthenium (μ-chloro)(μ-dihydrido) dimer was an efficient arene hydrogenation catalyst.[13,14] Again, η^4 coordination appears to be important in the postulated mechanism of arene ring hydrogenation with this system.[14a] Maitlis and co-workers found that the catalyst precursor with the formula [Cp*RhCl2]2 (Cp* = pentamethylcyclopentadienyl) was also efficient for arene hydrogenation in the presence of a base co-factor such as triethylamine.[15,16] The base is thought to be required to neutralize the HCl that is formed in the heterolytic cleavage of H_2 to form the active catalyst. High *cis* stereoselectivity was observed with disubstituted benzene derivatives and predominant formation of 1,2,3,4-tetrahydroanthracene from anthracene was also observed, again indicative of the importance of η^4 coordination.

Recently, Taube and co-workers discovered some interesting η^2 arene complexes, $(NH_3)_5Os(\eta^2-C_6H_6)$ that provided cyclohexene upon reaction with H_2 using a heterogeneous surface, Pd / C.[17,18] Other arene hydrogenation studies with purported homogeneous Rh complexes that included [(PhO)3P]2(AcAc)Rh(I)[19] and Rh amino acid complexes (anthranilic, N-phenylanthranilic)[20] were also published; the former paper did not provide enough information on catalyst stability and the latter complexes were not characterized structurally, therefore, making it difficult to fully understand this system. Larsen and co-workers have recently described a homogeneous ionic hydrogenation system that consisted of $H_2O \cdot BF_3$, H_2, and $(CH_3CN)_2PtCl_2$. This catalytic system reduced benzene, naphthalene, substituted naphthalenes, and other polynuclear arene derivatives. While mechanistic studies are still to be performed, it is possible that protonation provides a carbonium ion intermediate and this is trapped by a PtH complex. Thiols and thiophenols quench hydrogenation activity by presumably forming PtSR complexes that are inactive.[21]

The mechanisms of all these arene hydrogenation complexes depends on disruption of the aromaticity of the arene ring and it is obviously an area that needs further study. More examples of η^4 and η^2 arene complexes are necessary as is the study of these reactions using high pressure NMR techniques, which will be discussed in a subsequent section.

Hydrogenation of Mono and Polynuclear Heteroaromatic Nitrogen Compounds

Prior to 1981, relatively little was known about the regioselective reductions of heteroaromatic nitrogen compounds, even though this class of model coal compounds is highly important to study from a fundamental point of view as well as from the fact that these compounds are prevalent in petroleum products and are an economic and environmental concern.[5] The earlier studies were able to demonstrate that quinoline, a representative polynuclear heteroaromatic nitrogen model coal compound, could be reduced regioselectively to 1,2,3,4-tetrahydroquinoline (THQ) (eq 5) under strictly hydrogenation [H_2, RhCl$_2$Py$_2$(DMF)BH$_4$][22] and under sg conditions [CO, H_2, Mn$_2$(CO)$_8$(BuP)$_2$].[23]

$$(5)$$

As well, Laine and co-workers were able to demonstrate that Rh$_6$(CO)$_{16}$ under wgs conditions (CO, H_2O) catalyzed the reduction of pyridine to piperidine and other piperidine derivatives.[24]

This basically was the state of the art with regard to heteroaromatic nitrogen compound homogeneous hydrogenation chemistry and in 1981 Fish and co-workers initiated a program to elucidate the synthetic scope and the mechanism of the selective hydrogenation of polynuclear heteroaromatic nitrogen compounds.[5-7] The important discovery that under a wide variety of conditions regioselectivity of the nitrogen ring was maintained, i.e., wgs, sg, and H_2 alone, provided a common characteristic for all these homogeneous hydrogenation reactions; however, profound differences in mechanism prevailed with respect to the transition-metal catalyst. In addition, the polynuclear heteroaromatic nitrogen compounds were far more reactive under all the homogeneous hydrogenation conditions studied than their carbon analogues, eg, acridine > anthracene.[5,6]

The Mn, Fe, and Co carbonyl complexes studied under wgs or sg catalytic conditions with quinoline, 5,6- and 7.8-benzoquinolines, acridine, and phenathridine were found to proceed independently of CO pressure, i.e., binding to the metal center was not rate limiting.

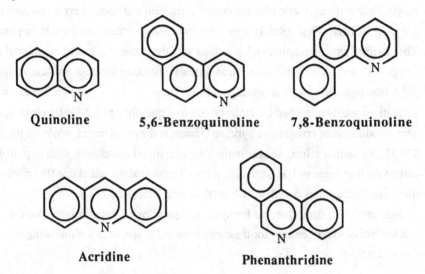

Quinoline   **5,6-Benzoquinoline**   **7,8-Benzoquinoline**

Acridine   **Phenanthridine**

Kaesz and co-workers did an extensive study of the $Fe(CO)_5$ catalyzed reductions of polynuclear heteroaromatic nitrogen compounds under wgs reaction conditions.[25] They found a similar regioselectivity for the nitrogen ring and speculated that $[HFe(CO)_4]^-$ is the active catalyst and that an electron transfer mechanism prevails. Alternatively, it was found that the Ru carbonyls, $Ru(Cl_2)(CO_2)(Ph_3P)_2$ and $H_4Ru(CO)_{12}$, were inactive in the presence of carbon monoxide; again, as with the polynuclear aromatic compounds, binding of the nitrogen compound to the metal center appears to be a pre-rate determining step.[5] Thus, CO was acting as a competitive inhibitor under wgs or sg conditions. Murahashi and co-workers studied the selective hydrogenation of a variety of substituted quinoline derivatives under wgs conditions with $Rh_6(CO)_{16}$ as the catalyst.[26] Apparently, this Rh carbonyl cluster catalyst is able to reduce the quinoline compounds in the presence of CO, unlike the Ru carbonyl cluster complexes, which are inhibited in the presence of CO. As well, they notice a change in regioselectivity when the hydrogenation reaction is carried out with H_2 alone at 150 °C; from 4-methyl-1,2,3,4-tetrahydroquinoline (wgs) to 4-methyl-5,6,7,8-tetrahydroquinoline. This may be due to Rh carbonyl cluster catalyst decomposition at 150 °C to Rh metal; the fate of this Rh complex at the reaction temperature required is not mentioned by these workers and is suspect.

In their quest for a more active homogeneous catalyst for selective hydrogenation of polynuclear heteroaromatic nitrogen compounds, Fish and co-workers discovered that $(Ph_3P)_3RhCl$ and $(Ph_3P)_3RuCl_2$ were excellent catalyst precursors for this purpose.[27-29] They defined structure-activity relationships with the above-designated model coal nitrogen compounds and found the order to be as follows: phenanthridine >> acridine >> quinoline > 5,6-benzoquinoline >> 7,8-benzoquinoline. They also found that pyridine and methyl derivatives totally quenched the hydrogenation of quinoline to 1,2,3,4-tetrahydroquinoline. This is indicative of competitive pyridine binding at the metal center, while the product THQ had a similar effect. Interestingly, they also found compounds such as pyrrole, carbazole, thiophene and, in particular, p-cresol enhanced the initial rate (% / min) of quinoline hydrogenation with the Rh catalyst by a factor of 1.5-2.5.

Substitution of deuterium gas for hydrogen gas allowed some insights into the mechanism of hydrogenation for these catalysts and is shown in the following scheme for the Rh catalyst.

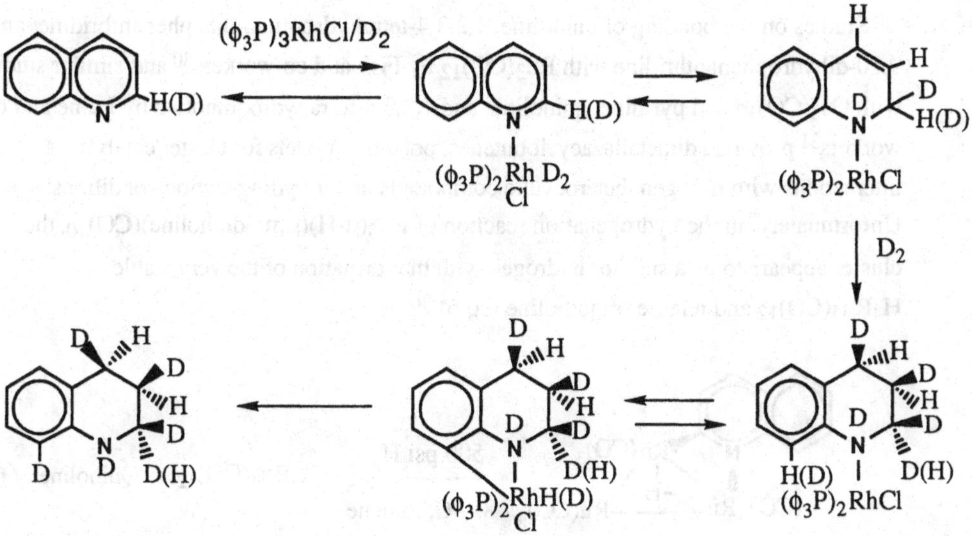

Several important observations concerning this mechanistic scheme are as follows: Quinoline and D_2 (H_2) binding to the Rh metal center provides the driving force for reduction of the C=N bond; a reversible process, which provides D for H exchange at the 2-position of quinoline. Reduction of the 3,4 double bond occurs with *cis* stereochemistry, while exchange of the aromatic hydrogen at the 8-position is a consequence of a cyclometallation reaction. This latter exchange at the 8-position from the reduced nitrogen ring intermediate was proven by using 1,2,3,4-tetrahydroquinoline as the starting substrate in the above scheme and finding that the 8-hydrogen was exchanged for deuterium; no other H exchange was found by NMR analysis.[27]

The differences between the Rh and Ru catalysts were pronounced: the initial rates of hydrogenation of the nitrogen substrates for Ru / Rh was a factor of 3; the Ru catalyst was inhibited by several nitrogen compounds as with the Rh catalyst, but no rate enhancement was observed as was shown for the Rh catalyst with the above-mentioned compounds; D for H exchange of positions 2 and 8 was found for 1,2,3,4-tetrahydroquinoline with the Ru catalyst.[28,29]

Studies on the bonding of quinoline, 1,2,3,4-tetrahydroquinoline, phenanthridine, and
9,10-dihydrophenanthridine with $Ru_3(CO)_{12}$ by Fish and co-workers[30] and similar studies
with $Os_3(CO)_{12}$ and pyridine, quinoline, and 1,2,3,4-tetrahydroquinoline by Laine and co-
workers[31] provided dimetallazacyclobutenes; potential models for cluster catalyst
interactions with nitrogen heterocyclic compounds under hydrogenation conditions.
Unfortunately, in the hydrogenation reaction of $Ru_3(\mu-H)(\mu,\eta^2$-quinoline$)(CO)_{10}$, the
cluster appears to be a sink for hydrogen with the formation of the very stable
$H_4Ru_4(CO)_{12}$ and release of quinoline (eq 6).[30]

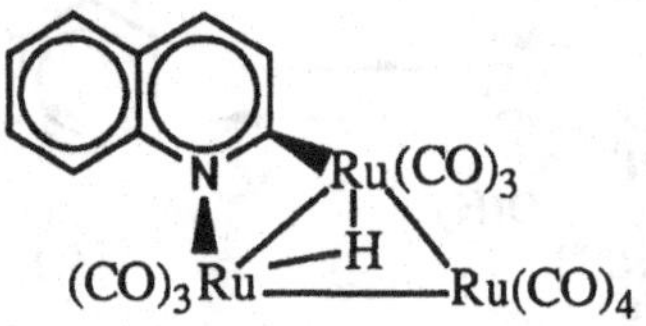

$$\xrightarrow[\text{85 °C, toluene}]{\text{500 psi } H_2} \quad H_4Ru_4(CO)_{12} + \text{Quinoline} \quad (6)$$

The importance of understanding the bonding mode of polynuclear heteroaromatic
nitrogen compounds to Rh metal centers can be directly related to the mechanism of
selective hydrogenation of these compounds. In order to provide critical information on
role of $N(\eta^1)$ or π (η^6)-bonding in the selective hydrogenation of quinoline and related
derivatives, Fish and co-workers initiated bonding and catalysis studies with
η^5-pentamethylcyclopentadienylrhodium dication derivatives ($Cp^*RhL_nX_2$, L = p-xylene,
$[CH_3CN]_3$, $[CH_3COCH_3]_3$; X= PF_6, BF_4) with a variety of mono and polynuclear
heteroaromatic nitrogen compounds.[32]

The reaction of quinoline with $Cp^*Rh(CH_3CN)_3{}^{2+}$ provided an air and moisture
sensitive complex, $Cp^*Rh(\eta^1$, N-quinoline$)(CH_3CN)_2{}^{2+}$:

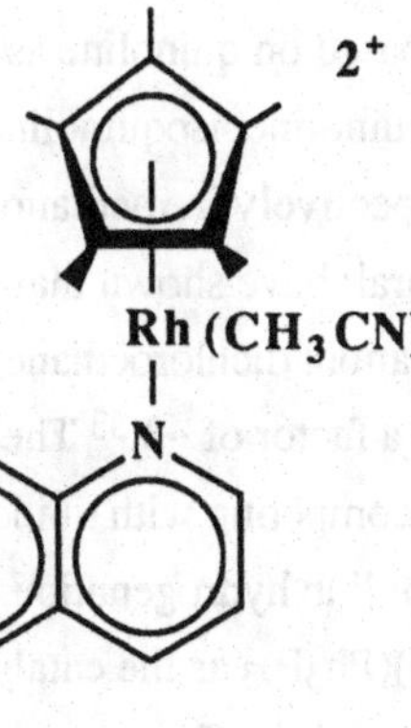

The reaction of this complex with traces of H_2O provided a crystalline derivative, $[Cp*Rh(\eta^1, N\text{-quinoline})(OH)]_2{}^{2+}$, whose single crystal X-ray structure unequivocally verified N-bonding to Rh. Alternatively, reaction of $Cp*Rh(CH_3CN)_3{}^{2+}$ with THQ gave the $Cp*Rh(\eta^6\text{-THQ})^{2+}$ complex:

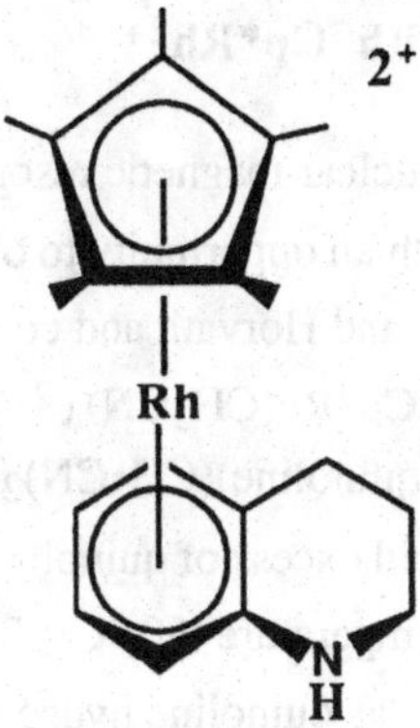

The significance of determining the N-bonding in the $Cp*Rh(\eta^1, N\text{-quinoline})(CH_3CN)_2{}^{2+}$ complex and the $\pi(\eta^6)$-bonding in $Cp*Rh(THQ)^{2+}$ was the fact that the synthetic precursor to these two complexes, $Cp*Rh(CH_3CN)_3{}^{2+}$, was an efficient catalyst or catalyst precursor for the selective hydrogenation reaction of quinoline to THQ; this represents the first example of an organorhodium cationic complex acting as a catalyst for this conversion.[32] Thus, it appears from this result that <u>N-bonding of the quinoline to the Rh metal center is necessary for selective hydrogenation of the nitrogen ring</u>. The relative rates of hydrogenation (based on quinoline as the standard with an initial rate of 1% / min) were determined for quinoline, isoquinoline, and 2-methylquinoline and were found to be 1, .03, and 0.43, respectively in methanol at 80 °C.

Recent results by Fish and Baralt have shown that solvent has a dramatic effect on the initial rate of quinoline hydrogenation; dichloroethane increasing the initial rate of quinoline hydrogenation over methanol by a factor of ~3.[33] The methanol apparently can reduce the quinoline hydrogenation rate by competing with quinoline for the Rh metal center. A similar solvent effect during quinoline hydrogenation was discovered by Sanchez-Delgado and co-workers with $HClRu(CO)(Ph_3P)_3$ as the catalyst and toluene, acetonitrile, and methanol as solvents. They compared the % conversion of quinoline to THQ for the three solvents and found that acetonitrile quenched the hydrogenation activity, while in toluene, the % conversion was 9 times that in methanol.[34]

High Pressure Nuclear Magnetic Resonance Studies: The Mechanism of Quinoline Hydrogenation with $Cp*Rh^{2+}$

The advent of high pressure nuclear magnetic resonance spectroscopy (HPNMR) provides the catalysis chemist with an opportunity to observe, in real-time, the catalytic reaction of interest.[35] Thus, Fish and Horvath and co-workers have studied the selective hydrogenation of quinoline with $Cp*Rh(CH_3CN)_3{}^{2+}$ as the catalyst using the HPNMR technique.[36] The $Cp*Rh(\eta^1, N\text{-quinoline})(CH_3CN)_2{}^{2+}$ complex, which is shown above, was prepared *in situ* with a 14-fold excess of quinoline in the sapphire HPNMR tube in methylene chloride-d_2 at probe temperature (30 °C). The NMR spectrum was run and the $Cp*Rh$ signal at 1.78 ppm as well as quinoline hydrogens from ~ 7.4-8.9 ppm were present. The HPNMR tube was then pressurized to 500 psi with D_2 gas and spectra were recorded at probe temperature for 23h.

The scheme shows the initial results of these HPNMR experiments and they are tentatively as follows: (1) the hydrogen on the 8-position of free quinoline at 8.2 ppm undergoes broadening, but no H / D exchange during nitrogen ring hydrogenation, while the signal for the starting complex at 1.78 ppm disappears with a concomitant formation of free acetonitrile and signals for new $Cp*Rh$ complexes between 0.7 and 1.5 ppm; (2) the hydrogen on the 2-position of free quinoline at 8.9 ppm is slowly exchanged in a reversible process as nitrogen ring reduction (1,2-N=C) proceeds; (3) the reversible reduction of the N=C bond places ~1.5 deuteriums at the 2-position of the product THQ; (4) we speculate that $Cp*Rh^{2+}$ migrates to the 3,4-double (η^2) from nitrogen, after C=N bond reduction, and reduces the olefin via addition of D_2; (5) the Cp*Rh then coordinates to the benzene ring (η^6), and this $Cp*Rh(\eta^6\text{-}THQ)^{2+}$ complex, as well as the above-mentioned Cp*Rh olefin hydrogenation product, possibly undergoes a ligand exchange with the excess free quinoline and D_2 gas to continue the catalytic cycle. Pertinently, no deuterium incorporation was observed on the aromatic ring of THQ.

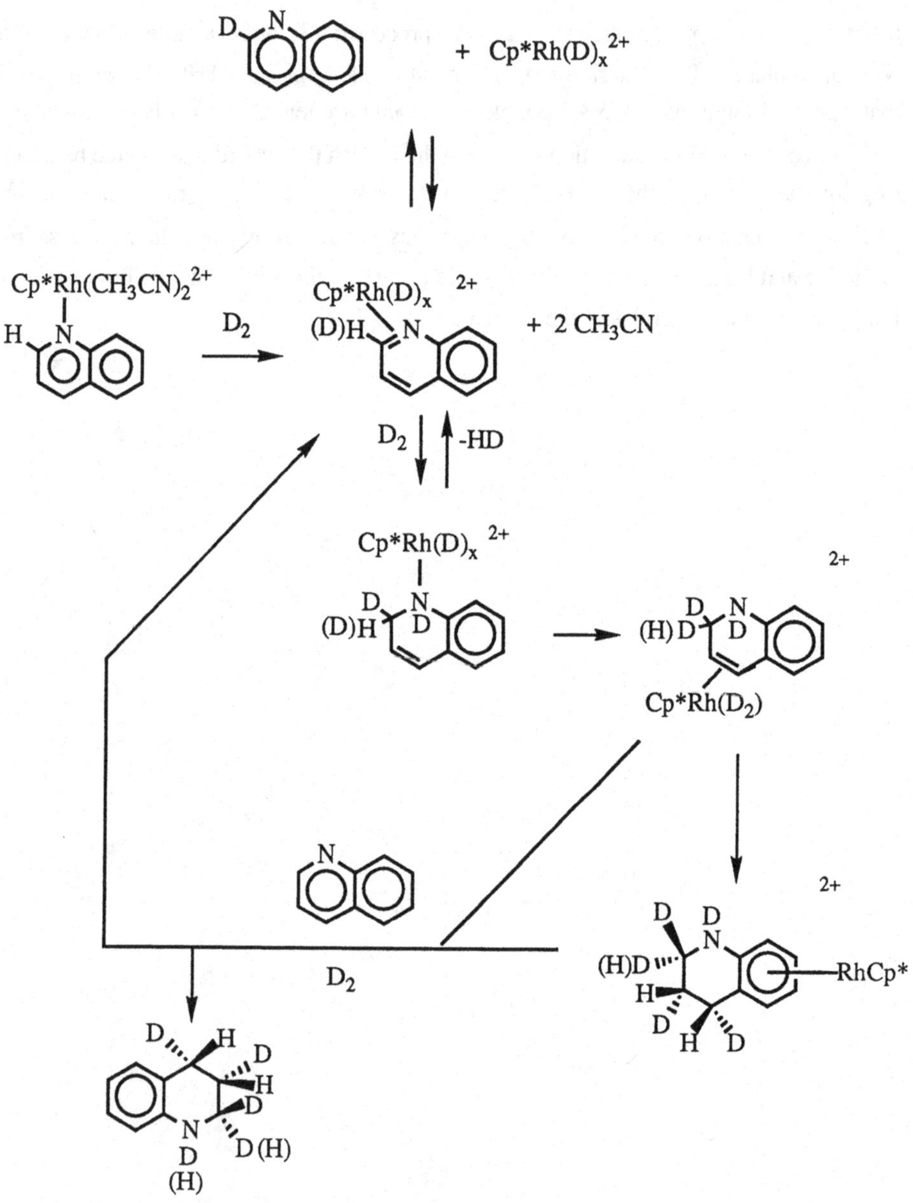
D
N
+ Cp*Rh(D)x 2+
Cp*Rh(CH3CN)2 2+
H
N
D2
Cp*Rh(D)x 2+
(D)H
N
+ 2 CH3CN
D2
-HD
Cp*Rh(D)x 2+
D
N
(D)H
D
2+
D
N
(H) D
D
Cp*Rh(D2)
N
D2
2+
D
D
N
(H)D
H
D
H
D
RhCp*
D
H
D
H
D
N
D
D (H)
(H)

The similarities to the above-mentioned proposed mechanism with $(Ph_3P)_3RhCl$ and its Ru analogue are apparent; however, there are significant differences such as the lack of aromatic H / D exchange. Clearly, this powerful HPNMR technique will allow more reliable information on homogeneous catalysis mechanisms than was previously known and further studies on other mono and polynuclear heteroaromatic nitrogen substrates are being carried out.

Conclusions

In this perusal of what is known about mono and polynuclear aromatic ring hydrogenation, it appears that new catalysts need to be discovered for these transformations and that more mechanistic information, particularly on the role of η^4 arene coordination during hydrogenation, is necessary. Studies on the regioselective hydrogenation of polynuclear heteroaromatic nitrogen compounds, under strictly H_2 conditions, has revealed that N-bonding to the organoRh and Ru centers is mandatory for selective hydrogenation to take place. The HPNMR technique will be able to answer questions about homogeneous catalysis mechanisms in real-time that will provide new insights into these interesting reactions. It is the hope of this author that this review will stimulate other workers in these areas to follow new directions.

Acknowledgment

The bonding and homogeneous catalysis studies on polynuclear heteroaromatic nitrogen compounds with transition-metal complexes that were carried out at LBL, as well as the preparation of this chapter, were generously supported by the Director, Office of Energy Research, Office of Basic Energy Sciences, Chemical Sciences Division of the U.S. Department of Energy under Contract No. DE-AC03-76-SF00098. Finally, the author would like to acknowledge his colleague, Heinz Heinemann of LBL, for stimulating his interest in this fascinating area of catalysis, and thank Martin Bennett of ANU for discussions on arene hydrogenation chemistry.

References

1. James, B. (1973) *Homogeneous Hydrogenation,* J. Wiley & Sons, New York

2. Friedman, S.; Metlin, S.; Svedi, A.; Wender, I. *J. Org. Chem.* **1959,** *24,* 1287.

3. (a) Feder, H. M.; Halpern, J. *J. Am. Chem. Soc.* **1975,** *97,* 7186. (b) Weil, T. A.; Friedman, S.; Wender, I. *J. Org. Chem.* **1974,** *39,* 48. (c) Taylor, P. D.; Orchin, M. *J. Org. Chem.,* **1972,** *37,* 3913.

4. Sweany, R.; Butler, S.C.; Halpern, J. *J. Organomet. Chem.* **1981,** *213,* 487.

5. Fish, R. H.; Thormodsen, A. D.; Cremer, G. A. *J. Am. Chem. Soc.* **1982,** *104,* 5234.

6. Fish, R. H. *Ann. N. Y. Acad. Sci.* **1983,** *415,* 292.

7. Cremer, G. A.; Vermeulen, T.; Fish, R. H. Homogeneous Hydrogenation of Model-Coal Compounds, LBL Report 14216, May 1982 (Ph.D. Thesis [in part] of G.A.C.).

8. (a) Halpern, J. *Pure & Appl. Chem.* **1987,** *59,* 173, and references therein. (b) Landis, C. R.; Halpern, J. *Organometallics* **1983,** *2,* 840.

9. Muetterties, E. L.; Bleeke, J. R. *Acc. Chem. Res.* **1979,** *12,* 324, and references therein.

10. Bleeke, J. R.; Muetterties, E. L. *J. Am. Chem. Soc.* **1981,** *103,* 556.

10a. Jonas, K. *Angew. Chem. Int. Ed. Engl.* **1985,** *24,* 295. The unpublished studies described were those of Jonas and Priemer (ref 85).

11. Johnson, J. W.; Muetterties, E. L. *J. Am. Chem. Soc.* **1977,** *99,* 7395.

12. Darensbourg, M. Y.; Muetterties, E. L. *J. Am. Chem. Soc.* **1978,** *100,* 7425.

13. Bennett, M. A.; Huang, T-N.; Turney, T. W. *J. Chem. Soc., Chem. Commun.* **1979,** 312.

14. Bennett, M. A. *Chemtech* **1980,** 444.

14a. In a private communication, Dr. Bennett has stated that the preparation of the arene hydrogenation catalyst, (hexamethylbenzene)ruthenium(μ-chloro)(μ-dihydrido) dimer, could not be repeated.

15. Russell, M. J.; White, C.; Maitlis, P. M. *J. Chem. Soc., Chem. Commun.* **1977,** 427.

16. Maitlis, P, M. *Acc. Chem. Res.* **1978,** *11,* 301, and references therein.

17. Harman, W. D.; Taube, H. *J. Am. Chem. Soc.* **1988,** *110,* 7906.

18. Harman, W. D.; Taube, H. *J. Am. Chem. Soc.* **1987,** *109,* 1883.

19. Pieta, D.; Trzeciak, A. M.; Ziolkowski, J. J. *J. Mol. Catal.* **1983,***18,* 193

20. Rajca, I. *Pol. J. Chem.* **1981,** *55,* 775.

21. Cheng, J. C.; Maioriello, J.; Larsen, J. W. *Energy & Fuels*, **1989**, *3*, 321.

22. Jardine, I.; McQuillin, F. J. *Chem. Commun.* **1970**, 626.

23. Derencsenyl, T. T.; Vermeulen, T. *Chem. Abst.* **1980**, *93*, 188 929. and LBL Report 9777, September 1979. (Ph.D. Thesis of T. T. D.).

24. Laine, R. M.; Thomas, D. W.; Cary, L. W. *J. Org. Chem.* **1979**, *44*, 4964.

25. Lynch, T. J.; Banah, M.; Kaesz, H. D.; Porter, C. R. *J. Org. Chem.* **1984**, *49*, 1266.

26. Murahashi, S-I.; Imada, Y.; Hirai, H. *Tetrahedron Letters*, **1987**, *28*, 77.

27. Fish, R. H.; Tan, J. L.; Thormodsen, A. D. *J. Org, Chem.* **1984**, *49*, 4500.

28. Fish, R. H. Tan, J. L.; Thormodsen, A. D. *Organometallics* **1985**, *4*, 1743.

29. Thormodsen, A. D.; Vermeulen, T.; Fish, R. H. LBL Report 22209, September, 1985 (Ph. D. Thesis of A.D.T.).

30. Fish, R. H.; Kim. T-J.; Stewart, J. L.; Bushweller, J.H.; Rosen, R. K.; Dupon, J. W. *Organometallics* **1986**, *5*, 2193.

31. Eisenstadt, A.; Giandomenico, C. M.; Frederick, M. F.; Laine, R. M. *Organometallics* **1984**, *4*, 2033, and references therein.

32. Fish, R. H.; Kim, H-S.; Babin, J. E.; Adams, R. D. *Organometallics* **1988**, *7*, 2250.

33. Fish, R. H.; Baralt, E. **1990** (manuscript in preparation).

34. Sanchez-Delgado, R. A.; Gonzalez, E. *Polyhedron* **1989**, *8*, 1431.

35. Roe, D. C. *J. Magn. Reson.* **1985**, *63*, 388.

36. Fish, R. H.; Baralt, E.; Kastrup, R. V.; Horvath, I. T. **1990** (manuscript in preparation).

SURFACE ORGANOMETALLIC CHEMISTRY ON OXIDES, ON ZEOLITES AND ON METALS

J.M. BASSET[*],   J.P. CANDY,   A. CHOPLIN,   M. LECONTE,
A. THEOLIER
Institut de Recherches sur la Catalyse
Conventionné avec l'Université Claude Bernard
2 avenue A. Einstein
69626 Villeurbanne Cedex
France

ABSTRACT. New concepts are slowly emerging from the overlap between organometallic chemistry and surface science. In particular the rules of molecular chemistry seem to apply quite well when organometallic complexes (mononuclear, polynuclear; transition metals, main group elements, lanthanides or actinides) react with a surface (oxide, zeolite, zerovalent metal). In the case of oxides it is thus possible to classify those surfaces by their chemical properties (e.g. acid-base, redox a.s.o.). Well defined surface organometallic fragments can also be prepared via organometallic complexes at the surface of oxides, zeolites and metals; it is thus possible to study on those fragments the real "elementary steps" of heterogeneous catalysis (oxidative addition, reductive elimination, insertion, C-C coupling, etc.). By means of supported heteropolynuclear complexes it is now possible to prepare bimetallic particles having the same composition as that of the starting complexes, a useful approach for the synthesis of tailor made catalysts. By means of organometallics it is possible to control the external pore size of zeolites and introduce a new approach of shape selectivation where the molecular steric control can be adjusted by the size of the organometallic fragment. Reaction of organometallics of main group elements with group VIII metals is a way to prepare well defined alloys of known composition which exhibit highly chemoselective properties in heterogeneous catalysis. It is also possible to graft organometallic fragments at the surface of metal particles and obtain a new generation of modified metallic catalysts extremely selective in a variety of catalytic reactions and which are already in practice in industry.

1.   Introduction

Surface organometallic chemistry (SOMC) is a relatively new field of chemistry which deals with the reactivity of organometallic compounds with surfaces. The approach is very general: the organometallic compounds considered are the main group elements complexes, the mononuclear complexes and clusters of the transition

85

R. Ugo (ed.), Aspects of Homogeneous Catalysis, Vol. 7, 85–115.
© 1990 Kluwer Academic Publishers. Printed in the Netherlands.

metals (in different oxidation states), the complexes of lanthanides and actinides a.s.o. The surfaces can be the surface of the "oxides", the surface (internal or external) of zeolites or the surface of a supported noble metal particle in a zerovalent or in a higher oxidation state. One can see easily that this field of chemistry which concerns primarily catalysis but also to a smaller extent surface science, material science a.s.o., is very large. New chemical concepts are slowly emerging. These concepts are first mechanistic ones: when an organometallic compound reacts with a surface, the observed chemistry is governed by the rules of molecular chemistry which seem to apply quite well to surfaces. To a certain extent the rules of surface science are also followed but these rules are probably less determining to explain surface chemistry. Secondly these concepts are practical ones: one can prepare, via an organometallic compound, the well defined surface "sites" that are mostly lacking in heterogeneous catalysis. The present article which tries to be comprehensive is not exhaustive; it is obviously restricted to a limited number of examples and reflects the point of view of the authors on a field which has developped tremendously in the last decade.

2.   Basic Rules Governing The Reactivity Of Organometallic Compounds With Surfaces Of Inorganic Oxides

Although this field is still in its infancy, a great variety of surface reactions leading to well defined surface complexes are already known (ref. 1). All these reactions seem to obey the same general rules (elementary steps) as those already observed in molecular chemistry. In the following examples, the surface of inorganic oxides behaves as a common "reagent" of molecular chemistry. Apparently, the reactions which occur on a surface may be described in the same way as any reaction occuring in solution. Obviously differences are also likely to occur due to steric crowding and rigidity of the "surface ligand". The purpose of this comprehensive approach will be to focus mostly on the analogies which exist rather than on possible differences.

2.1.   NUCLEOPHILIC ATTACK AT COORDINATED CO

On basic oxides such as partially hydroxylated magnesia, alumina, zinc oxide or lanthanum oxide, group VIII metal carbonyls such as $M_3(CO)_{12}$, M = Fe, Ru, Os, have a tendency to undergo a nucleophilic attack by surface hydroxyl groups (ref. 3). This phenomenon results in the formation of the corresponding anionic hydrides. The elementary steps leading to the formation of these anions are not yet fully understood but a reasonable path deduced from what is proposed in molecular chemistry (ref. 3) is probably the following:

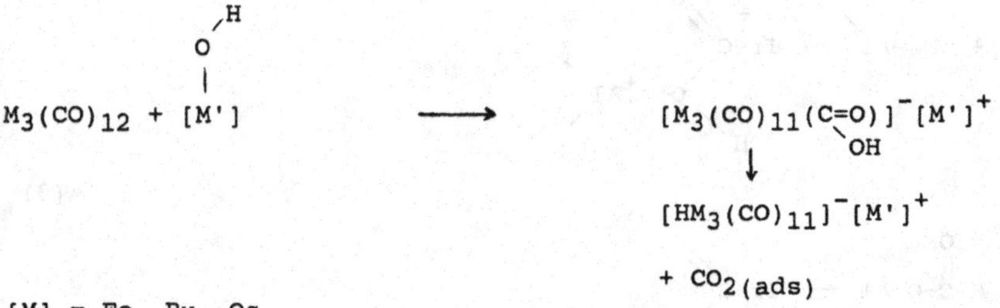

$$M_3(CO)_{12} + [M'] \longrightarrow [M_3(CO)_{11}(C\!\!=\!\!O)]^- [M']^+ \underset{OH}{}$$

$$\downarrow$$

$$[HM_3(CO)_{11}]^- [M']^+$$

$$+ \ CO_{2\,(ads)} \tag{1}$$

[M] = Fe, Ru, Os

[M'] = Mg, Zn, Al, La

    This nucleophilic character of surface OH groups towards coordinated CO seems to vary according to the support. The following order has been observed: Mg-OH > Zn-OH > La-OH > Al-OH. It is probably involved in some catalytic reactions which occur on surfaces between carbon monoxide and water. The water gas shift reaction could very likely obey some of these elementary steps:

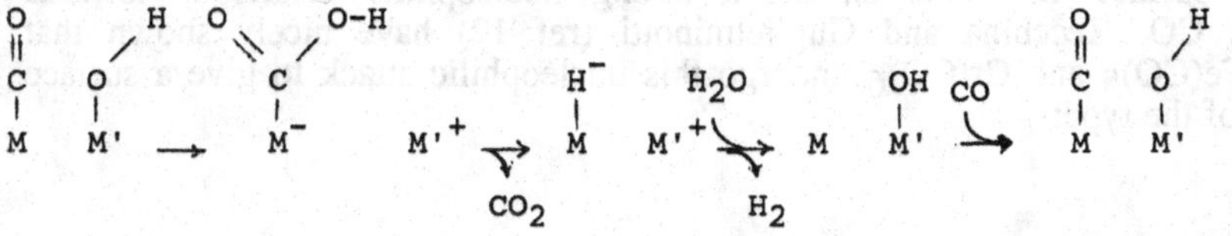

$$\tag{2}$$

    In some cases, the reactivity of carbonyl ligands is slightly different from that observed previously. When $Ir_4(CO)_{12}$ is chemisorbed on alumina, one observes the formation of a surface formate and simultaneously the cluster is transformed into very small particles of iridium covered with CO, (ref. 11a). The mechanism that has been proposed is slightly different from the previous one. One can assume first a nucleophilic attack at coordinated CO, as in the previous case, but then the ß-H elimination would be concerted with the coordination of the oxygen atom of the carboxylic ligand to the exposed aluminium ion:

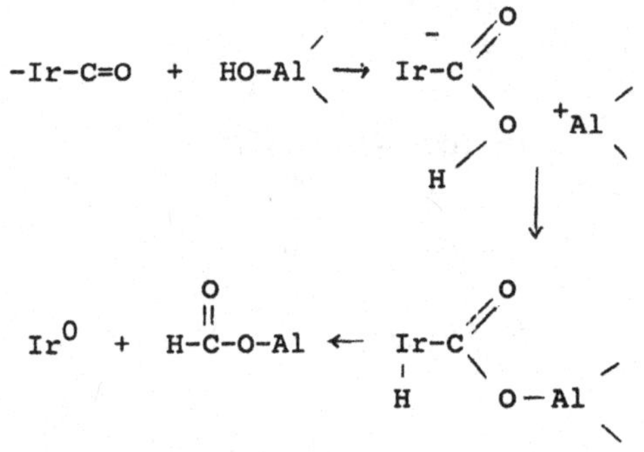

$$-Ir-C\equiv O \;+\; HO-Al\overset{/}{\underset{\backslash}{}} \;\longrightarrow\; Ir-C\overset{-}{\underset{\substack{O \\ | \\ H}}{\overset{/\!\!/\,O}{\diagdown}}} \; {}^+Al\overset{/}{\underset{\backslash}{}}$$

$$\downarrow$$

$$(3)$$

$$Ir^0 \;+\; \underset{\substack{| \\ H}}{\overset{\substack{O \\ || }}{H-C}}-O-Al \;\longleftarrow\; Ir-C\overset{/\!\!/\,O}{\underset{O-Al\diagup}{\diagdown}}$$

Formally, this mechanism explains how carbon monoxide can be inserted into the O-H bond of alumina, a reaction which is known to be catalyzed by zerovalent metals at the surface of basic oxides (ref. 11b, 11c):

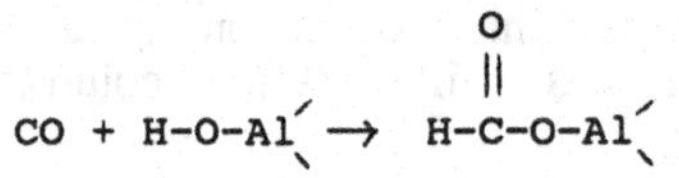

$$CO + H-O-Al\overset{/}{\underset{\backslash}{}} \;\longrightarrow\; \overset{\substack{O \\ ||}}{H-C}-O-Al\overset{/}{\underset{\backslash}{}} \qquad\qquad (4)$$

When the basic oxides are fully dehydroxylated to give a "pure" inorganic oxide, the surface O^{2-} ions exhibit a strong nucleophilic behaviour towards coordinated CO. Zecchina and Guglielminotti (ref. 12) have nicely shown that $Ni(CO)_4$, $Fe(CO)_5$ and $Cr(CO)_6$ undergo this nucleophilic attack to give a surface compound of the type:

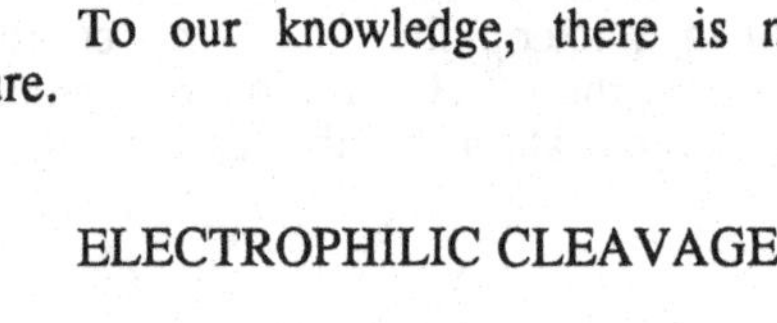

$$\underset{Mg^{2+}}{\overset{ML_n}{\underset{O}{\overset{|}{\underset{\diagup\diagdown}{C}}}\;\;O^{2-}}} \qquad\qquad \underset{Mg}{\overset{ML_n}{\underset{O\;\;\;\;O}{\overset{|}{\underset{\diagdown\,\diagup}{C}}}}} \qquad\qquad\qquad \begin{aligned} ML_n &= Fe(CO)_4 \\ &\quad\; Ni(CO)_3 \\ &\quad\; Cr(CO)_5 \end{aligned}$$

To our knowledge, there is no molecular analogue of such surface structure.

2.2. ELECTROPHILIC CLEAVAGE OF METAL-CARBON BONDS

This kind of reaction occurs when a metal-alkyl or metal-allyl bond is allowed to react with the surface of an oxide which has electrophilic OH groups. It is is mostly observed on silica, titania, and sometimes partially dehydroxylated alumina. A typical example is the reaction of a partially dehydroxylated silica with $Rh(\eta^3\text{-}C_3H_5)_3$ which has been studied recently in various laboratories (refs. 5, 12, 13):

$$\begin{array}{c} \diagdown\\ \text{Si-OH}\\ \diagup\\ \text{O}\\ \diagdown\\ \text{Si-OH}\\ \diagup \end{array} + Rh^{III}(\eta^3\text{-}C_3H_5)_3 \rightarrow \begin{array}{c} \diagdown\\ \text{Si-O}\\ \diagup\\ \text{O} \qquad Rh^{III}\\ \diagdown\\ \text{Si-O}\\ \diagup\\ \text{H} \end{array} + \quad =\diagup \qquad (5)$$

$$\begin{array}{c} \diagdown\\ \text{Ti-OH}\\ \diagup\\ \text{O}\\ \diagdown\\ \text{Ti-O}\\ \diagup\ |\\ \text{Ti}\\ \diagup \end{array} + Rh^{III}(\eta^3\text{-}C_3H_5)_3 \rightarrow \begin{array}{c} \diagdown\\ \text{Ti-O}\\ \diagup\\ \text{O} \qquad Rh^{III}\\ \diagdown\\ \text{Ti-O}\\ \diagup\ |\\ \text{Ti}\\ \diagup \end{array} + \quad =\diagup \qquad (6)$$

Such reaction does not occur with a fully hydroxylated alumina nor with hydroxylated magnesia (ref. 13). It involves very likely acidic OH groups which make an electrophilic cleavage at metal-carbon bond:

$$-\text{Si-OH} + Rh \diagdown\!\!\diagup\!\!\backslash\backslash \rightarrow -\text{Si-O} \overset{Rh}{\diagup}\!\!\diagdown\!\!\diagup\!\!\backslash\backslash \rightarrow -\text{Si-O} \overset{Rh}{\diagup} + \; =\diagup \qquad (7)$$

Similar reactions seem to occur when alkyl groups coordinated to highly oxophilic metals react with silica (refs. 5c, 5d):

$$\begin{array}{c} \diagdown\\ \text{Si-OH}\\ \diagup\\ \text{O}\\ \diagdown\\ \text{Si-OH}\\ \diagup \end{array} + M(R)_4 \longrightarrow \begin{array}{c} \diagdown\\ \text{Si-O} \qquad R\\ \diagup\qquad\diagup\\ \text{O}\qquad M^{IV}\\ \diagdown\qquad\diagdown\\ \text{Si-O}\qquad R\\ \diagup \end{array} + 2RH \qquad (8)$$

M = Ti, Zr; R = alkyl, benzyl

An important class of polymerization catalysts have been prepared in the past following similar types of reactions using chromocene as the precursor (ref. 15):

$$(\eta^5\text{-}C_5H_5)_2Cr \quad + \quad \overset{\backslash}{\underset{/}{-Si\text{-}OH}} \quad\longrightarrow\quad \overset{\backslash}{\underset{/}{-Si\text{-}O\text{-}Cr}}(\eta^5\text{-}C_5H_5) + C_5H_6$$

$$(\eta^5\text{-}C_5H_5)_2Cr \quad + \quad O\begin{array}{c} \diagup \,Si\text{-}OH \\ \diagdown \\ \diagdown\, Si\text{-}OH \end{array} \quad\longrightarrow\quad O\begin{array}{c} \diagup Si\text{-}O \diagdown \\ \diagdown \quad\quad Cr^{II} \\ \diagdown Si\text{-}O \diagup \end{array} + \; 2C_5H_6$$

There appears to be two sequential hydrolysis processes on silica. Higher degree of hydroxylation allows the presence of adjacent hydroxyl sites and the hydrolysis of the two Cp ligands becomes possible.

Metal alkyls of main group elements also react with acidic OH groups of silica probably by electrophilic cleavage of the metal carbon bond. For example $Sn(n\text{-}C_4H_9)_4$ reacts with a partially dehydroxylated silica $[(SiO_2(200)]$ according to the following reaction (ref. 16):

$$O\begin{array}{c} \diagup Si\text{-}OD \\ \diagdown \\ \diagdown\, Si\text{-}OD \end{array} \quad + \quad Sn(n\text{-}C_4H_9)_4 \underset{T>150°C}{\longrightarrow} 2\; C_4H_9D \; + \; O\begin{array}{c} \diagup Si\text{-}O \diagdown \quad \diagup C_4H_9 \\ \quad\quad Sn \\ \diagdown Si\text{-}O \diagup \quad \diagdown C_4H_9 \end{array}$$

The number of leaving alkyl groups seems to depend upon the degree of hydroxylation of silica.

$MgNp_2$ also reacts with a partially dehydroxylated silica with formation of a magnesium oxygen bond (refs. 17, 18):

$$\overset{\backslash}{\underset{/}{-Si\text{-}OH}} + MgNp_2 \quad\longrightarrow\quad \overset{\backslash}{\underset{/}{-Si\text{-}O\text{-}Mg}} \!\!-\!\!\times \quad\quad + \; NpH$$

This surface species is thermally stable as it is expected with compounds of the type $Mg(R)(OR)$ (ref. 19).

LiNp also reacts with silanol groups of silica to give a lithiated surface according to the reaction (ref. 18):

$$\overset{\backslash}{\underset{/}{-Si\text{-}OH}} + LiNp \quad\longrightarrow\quad \overset{\backslash}{\underset{/}{-Si\text{-}O\text{-}Li}} + NpH$$

One should remark that the electrophilic cleavage of metal-carbon bonds which seems to occur in several cases on partially dehydroxylated oxides is not necessarily the only pathway to cleave metal-alkyl groups. For example it is known that $Cp_2\,MMe_2$ (M = Th, U) undergoes a ready protonolysis in solution:

$$Cp_2^*M(CH_3)_2 \ + \ ROH \ \rightarrow \ Cp_2^*M(OR)(CH_3) \ + \ CH_4$$

$$Cp_2^*M(OR)(CH_3) \ + \ ROH \ \rightarrow \ Cp_2^*M(OR)_2 \ + \ CH_4$$

      This might represent effectively a good model for an electrophilic cleavage reaction at a surface hydroxyl group. However careful ^2H labeling experiments demonstrated that the equivalent surface reaction was only one of the three alternative mechanisms leading to the elimination of methane (ref. 20):

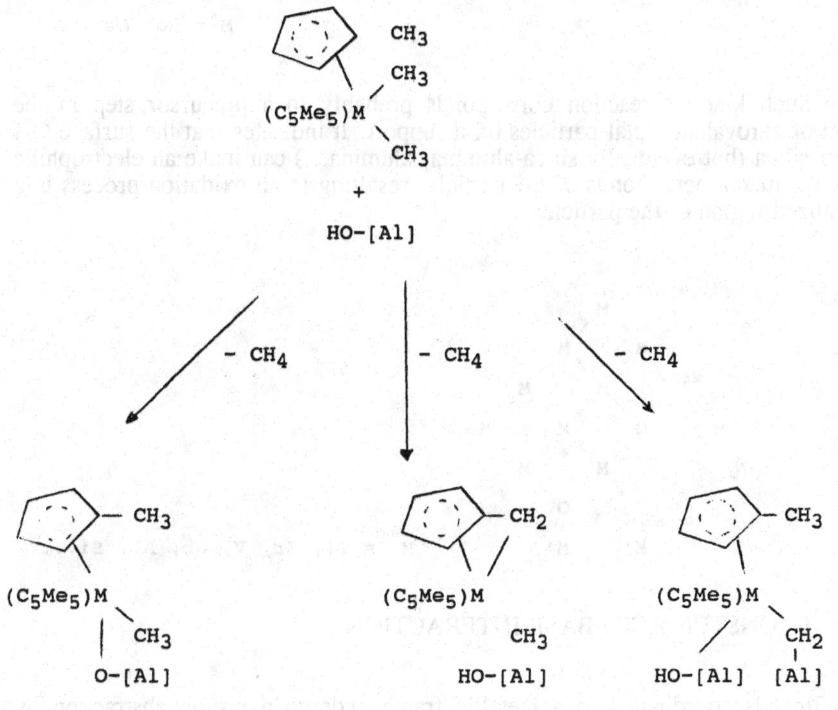

2.3. OXIDATIVE ADDITION OF SURFACE O-H GROUPS TO ZEROVALENT METALS

When zerovalent metal clusters such as $Ru_3(CO)_{12}$ or $Os_3(CO)_{12}$ react with an oxide surface covered with slightly acidic OH groups such as silanols, an oxidative addition occurs with formation of an hydrido cluster (ref. 6):

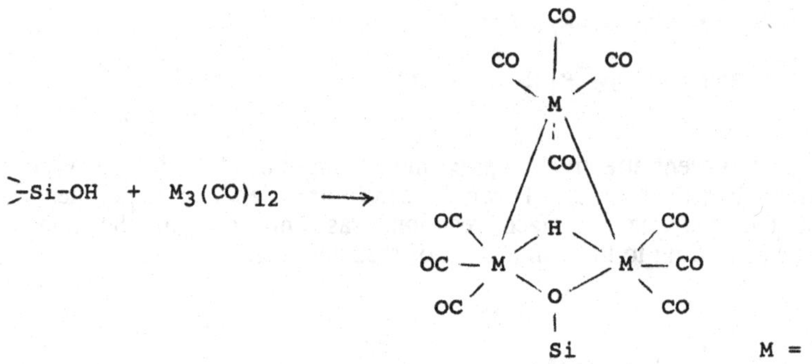

$$\gtrsim\!Si\text{-}OH \ + \ M_3(CO)_{12} \ \longrightarrow \qquad\qquad\qquad + \ 2 \ CO$$

$$(10)$$

$$M \ = \ Ru, \ Os$$

Such kind of reaction corresponds probably to a precursor step in the oxidation of zerovalent metal particles on a support. It indicates that the surface OH groups of silica (but eventually silica-alumina, alumina...) can make an electrophilic attack at the metal-metal bonds of the particle, resulting in an oxidation process in a very localized region of the particle:

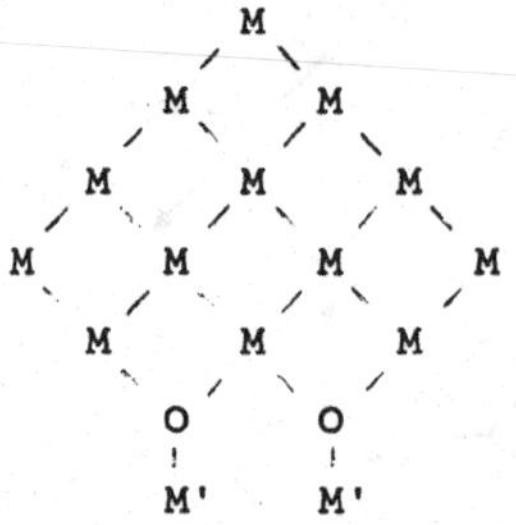

$$M' \ = \ Ti, \ Zr, \ V, \ Nb, \ Al, \ Si...$$

2.4. BRÖNSTED ACID-BASE INTERACTION

Hydrido ligands coordinated to a metallic frame undergo a proton abstraction by surfaces of basic oxides, such as alumina and magnesia, leaving a negative charge on the complex (refs. 7, 21):

$$L_nM\text{-}H \ + \ Mg\text{-}OH \ \rightarrow \ [\ Mg(OH_2) \ ]^+[L_nM]^- \qquad\qquad (11)$$

$$L_nM_x\text{-}M_y'\text{-}H \ + \ Mg\text{-}OH \ \rightarrow \ [\ Mg(OH_2) \ ]^+[L_nM_xM_y']^- \qquad (12)$$

This acid-base behaviour has been observed with mononuclear (ref. 7) and polynuclear complexes (ref. 21).

The fact that polynuclear hydrido clusters undergo easily a proton abstraction on basic oxides may be at the origin of new concepts in catalysis to explain some aspects of "metal support interaction". It is well known for example that the ν(CO) frequency at full coverage of CO chemisorbed on supported rhodium depends quite strongly on the basicity of the support (ref. 8a). It may be possible that during the high temperature reduction process of metallic salts under H_2, the surface hydrogen adsorbed on small particles migrates on the support under the form of a proton. The resulting small particle might develop a slight negative charge due to this spillover of protons.

2.5.   LEWIS ACID-BASE INTERACTION

Carbonyl ligands coordinated to a metallic frame undergo an acid-base interaction with the Lewis centers of a surface. Apparently, this phenomenon may occur with terminal and μ_{-2} bridging CO (refs. 3, 10, 22) :

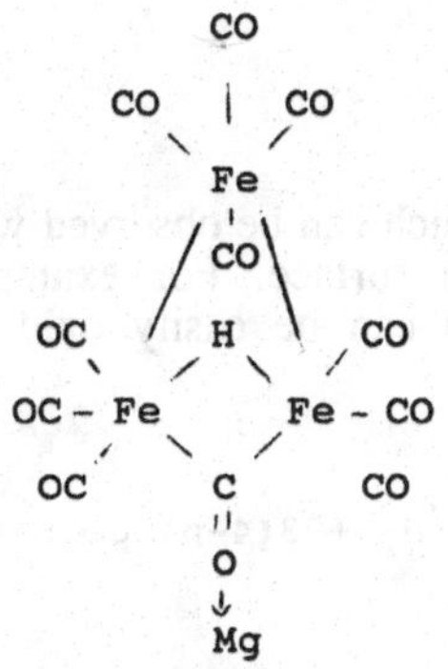

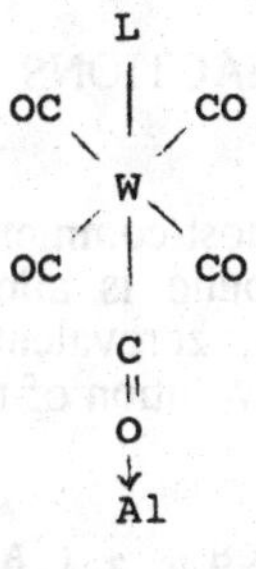

The implications, in catalysis, of such electrophilic attack at coordinated CO are probably very wide. Easy formation of C-C bond by CO insertion into a metal alkyl bond may thus be favored. However, in order to really prove the occurence of such phenomena in catalysis, it is necessary to study the elementary steps simultaneously on surfaces and on molecular models. In this respect it has been nicely shown by Shriver (ref. 10) that Lewis acid-Lewis base interaction can occur on the same complex/support system and promote the insertion of CO into a metal-alkyl bond:

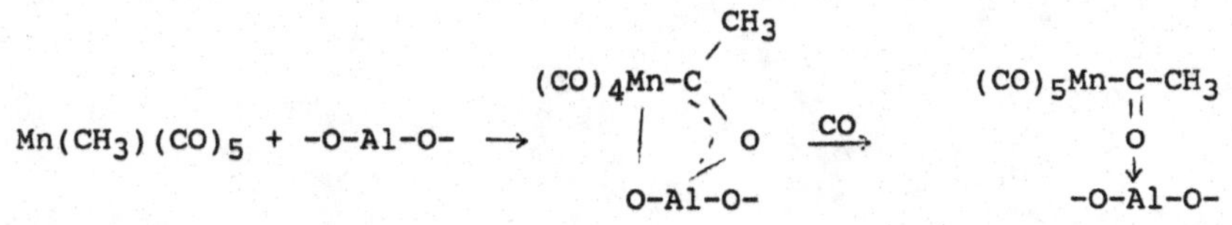

The same kind of reaction (ref. 1O) was observed when $Mn(CH_3)(CO)_5$ was allowed to react with $AlBr_3$ in the presence of CO, showing that surface chemistry strongly parallels molecular chemistry in solution:

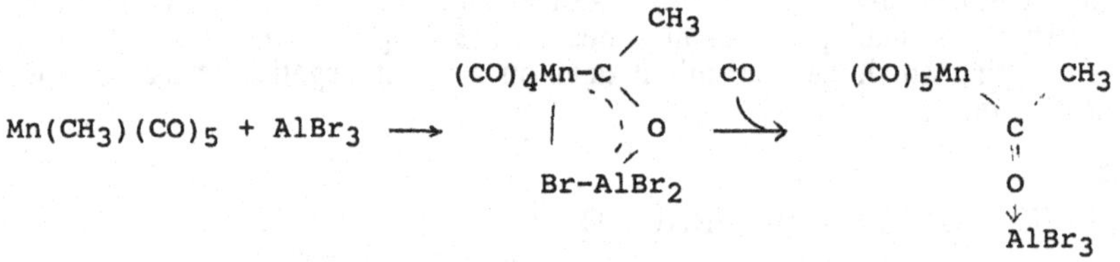

$$Mn(CH_3)(CO)_5 + AlBr_3 \longrightarrow \quad \begin{array}{c} CH_3 \\ (CO)_4Mn-C \\ | \quad \cdot \, O \\ Br-AlBr_2 \end{array} \quad CO \longrightarrow \quad \begin{array}{c} (CO)_5Mn \quad CH_3 \\ C \\ \| \\ O \\ \downarrow \\ AlBr_3 \end{array}$$

It has been suggested that the role of alkali "additives" in Fischer-Tropsch synthesis was to promote the CO insertion into a surface metal-alkyl bond (ref. 23). This question remains open and is still a matter of strong interest at the moment.

2.6.   REDOX REACTIONS

This is probably the most common class of reactions which can be observed when an organometallic compound is allowed to react with a surface. For example, on hydroxylated alumina, zerovalent carbonyl complexes can be easily oxidized by surface protons with evolution of molecular hydrogen:

$$Os_3(CO)_{12} + 6 \, Al-OH \rightarrow 3 \, (Al-O)_2Os^{II}(CO)_n + 3 \, H_2 + 3(4-n) \, CO \quad n = 2, \, 3$$

$$Rh_6(CO)_{16} + 6 \, Al-OH \rightarrow 6 \, (Al-O)(Al-OH)Rh^{I}(CO)_2 + 3 \, H_2 + 4 \, CO$$

This kind of redox reaction will govern the oxidation state of a given metal on a given support. A catalytic reaction may result from the redox properties of the metal/support system. One typical example is the reactivity of $Rh_6(CO)_{16}$ with surface OH groups on alumina which leads to the oxidation of zerovalent rhodium to a rhodium(I) dicarbonyl surface complex with simultaneous evolution of H_2. However in the presence of CO and water, reduction of rhodium(I) to Rh(0) occurs so that during catalysis one observes on the surface, both the cluster and the Rh(I) dicarbonyl. A tentative scheme for the mechanism of this reaction is given below (ref. 8c):

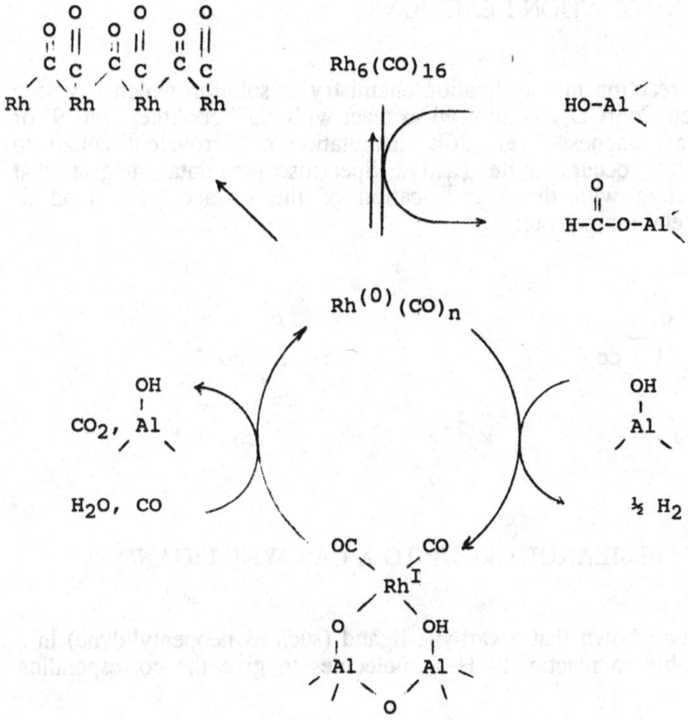

It should be noted here that the fact that $Rh_6(CO)_{16}$ is observed during the catalytic phenomenon is not a proof of its participation to the catalytic cycle: in this respect it has been found that, depending on CO and H_2O pressures, zerovalent rhodium can exist on the surface as metallic particles or as $Rh_6(CO)_{16}$ (ref. 8c). One should also notice that the OH groups of alumina can behave as nucleophiles towards CO coordinated to the cluster frame with the formation of surface formate, or as electrophile towards zerovalent rhodium with formation of Rh(I) and hydrogen (redox process).

One should also notice that the chemistry which is derived from the redox behaviour of the Rh/Al-OH system is not limited to supported organometallics. It is more than likely that the behaviour of metallic rhodium supported on alumina obeys similar or even identical rules. Recent works (ref. 24) related on the topic of Rh/Al_2O_3 systems have lead to the conclusion that oxidation of Rh is achieved by surface protons rather than by a dissociative chemisorption of CO on small rhodium particles (ref. 25).

2.7.   DISPROPORTIONATION REACTIONS

This is a well known reaction in coordination chemistry in solution which has been observed recently when $Co_2(CO)_8$ is allowed to react with NaY zeolithes (ref. 9) or a basic oxide such as magnesia (ref. 26): dismutation of zerovalent cobalt to $[Co(CO)_4]^-$ and Co^{2+} occurs quite easily. Spectroscopic data suggest that $[Co(CO)4]^-$ is interacting with the Mg^{2+} cation of the surface by a kind of heteronuclear metal-metal ion pairing:

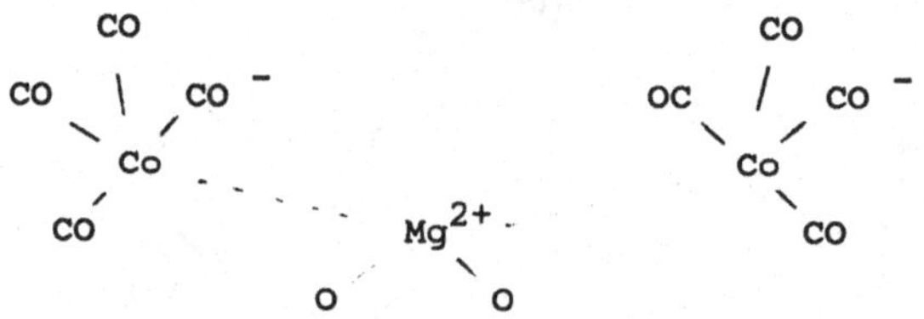

2.8.   ADDITION OF SILANOL GROUP TO A CARBYNE LIGAND

Schrock et al. (27) have shown that a carbyne ligand (such as neopentylidyne) in a tungsten complex is able to react with H-Y molecules to give the corresponding carbene complex:

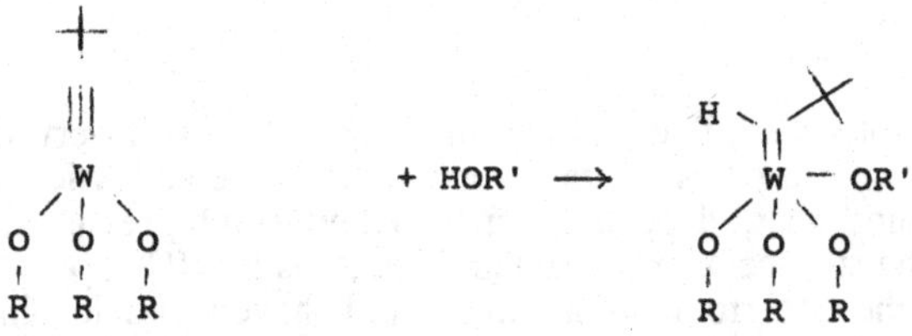

If the carbyne complex is allowed to react with surface silanol groups of silica, a similar reaction occurs with formation of a relatively well defined surface carbene complex which is active for the olefin metathesis reaction (ref. 28):

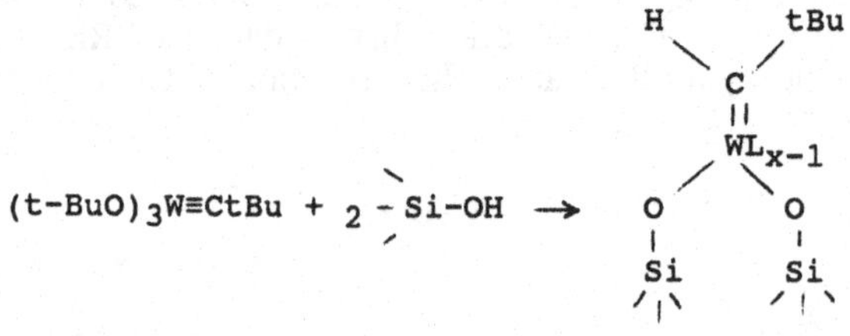

$L = tBuO$

3. **Some Elementary Steps in Heterogeneous Catalysis from Well Defined Surface Organometallic Fragments**

The knowledge of the mechanism of action of many homogeneous catalysts has been growing rather quickly in the last decades. This fact is mainly due to a fast development of molecular chemistry especially organometallic chemistry which allowed the proposal of a variety of elementary steps. In contrast, progresses in the knowledge of the mechanism of action of most solid catalysts still remain more limited. In many cases, the nature of the catalyst-reactant interaction remains uncertain and the elementary steps are not precisely well known. In a few cases, detailed mechanisms can be advanced but those examples are modest in comparison with structural and mechanistic details that have been developed in molecular chemistry (ref. 1).

One of the difficulties encountered in heterogeneous catalysis is that the surfaces of solid catalysts are extremely complicated. The number of the so-called "active sites" can be very small with regards to the overall surface and consequently their structure is almost unknown at an atomic level. It is probably this complexity which has been at the origin of the development of surface organometallic chemistry: one may hope that the results in this later area will help the understanding of heterogeneous catalysis in the same way as organometallic chemistry has been at the origin of a better understanding of the mechanism of action of homogeneous catalysts. One of the reasons for such growing interest in surface organometallic chemistry is probably related with the organometallic character of a working site in surface catalysis.

In a first approximation, one can consider that during a catalytic cycle, the molecule(s) which interact(s) with the surface make(s), at least during one elementary step, one or several chemical bonds with one or several surface atoms. Therefore the so-called "active site", during the catalytic process, is a kind of supramolecule which includes both a fragment of the molecule(s) and one or a few atoms from the surface. If the substrate is an organic molecule, the "working site" has therefore a "surface organometallic character" and hopefully the rules of organometallic chemistry can be applied to this supramolecule:

$$Y - X - M - X - Y$$

It is therefore temptating to study the <u>elementary steps</u> of heterogeneous catalysis on these "<u>well defined supramolecules</u>" which can be considered as intermediates in surface reactions. In order to achieve this goal, it is necessary to synthesize such surface organometallic entities. As usual, on surfaces, it is difficult to be completely confident in the proposal of surface structure and surface reactivity. Consequently, the synthesis of molecular models of these supramolecules constitutes a necessary step in a molecular approach of surface catalysis. The examples that will be presented in this paragraph represent a few experiments carried out with some well defined surface organometallic fragments. The conclusions that can be drawn are still very limited.

3.1.   REVERSIBLE DISSOCIATION OF A MOLECULAR LIGAND

Many groups have reported the synthesis of alumina supported $Rh^I(CO)_2$ (ref. 8). For example, it is possible to oxidize $Rh_6(CO)_{16}$ to $Rh^I(CO)_2$ by reaction of hydroxyl groups of alumina with the carbonyl cluster (vide supra). Interestingly the carbonyl ligands of the mononuclear $Rh^I(CO)_2$ can be thermally dissociated, without further oxidation of Rh(I) by the remaining surface OH groups (refs. 8c, 8d):

$$({}_{/}^{\backslash}Al-O)({}_{/}^{\backslash}Al-OH)Rh(CO)_2 \rightleftharpoons ({}_{/}^{\backslash}Al-O)({}_{/}^{\backslash}Al-OH)Rh^I + {}_2 CO$$

One should mention here that such an equilibrium could never be observed in solution probably because it would occur at a temperature where a $Rh^I(CO)_2$ analogue such as $[Rh(CO)_2Cl]_2$ would not be stable. In any case, stable coordinative unsaturation is more easily observed on surface of solids than in the molecular state.

3.2.   "MOBILITY" OF A CLUSTER WITH RESPECT TO SURFACE OXYGEN ATOMS

The cluster $(\mu-H)(\mu-OSi\overset{\frown}{=})Os_3(CO)_{10}$ is coordinated to a silica surface via a μ-2 oxygen atom, which thus behaves as a 3 electron donor ligand. This cluster does not coordinate ethylene at room temperature (refs. 8d, 29) (Fig. 1). At 80° C, about 1 mole of ethylene can be reversibly coordinated to the grafted cluster. A model reaction, carried out with $(\mu-H)(\mu-OPh)Os_3(CO)_{10}$, indicates that this reaction occurs by opening of the μ_2O ligand into a terminal O ligand. This phenomenon indicates how mobile are transition metal atoms towards surface oxygen atoms whether this transition metal is in a zerovalent oxidation state or not. Perhaps more important is the fact that in a first approximation, one can ascribe to surface oxygen atoms a number of electrons which can govern surface ligand arrangements of metal atoms. As a consequence the elementary rules of coordination chemistry can be tentatively applied to all the elementary steps of a surface reaction in which the surface behaves as a ligand. Simple rules such as the 14-16-18 electron rule, applied to the surface organometallic entity might then help to predict the various steps of a catalytic cycle (ref. 2a). If this concept is valid when a complex is coordinated to the surface of an oxide, maybe it could also apply when the metal is present in the first layer of the lattice.

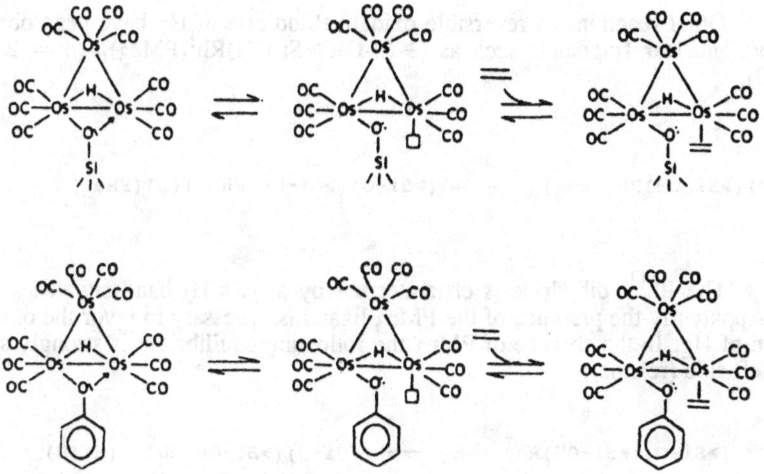

Fig. 1. Reversible coordination of ethylene to the supported cluster (μ-H) (μ-OSi≤)Os$_3$(CO)$_{10}$ and its molecular analogue (μ-H)(μ-OPh)Os$_3$(CO)$_{10}$ according to refs. 8d and 19.

3.3.   REVERSIBLE OXIDATIVE ADDITION OF H$_2$

The surface cluster (μ-H)(μ-OSi≤)Os$_3$(CO)$_{10}$ is able to react with H$_2$ at 100°C to give a surface cluster which is probably H$_2$Os$_3$(CO)$_{10}$(≥Si-OH) (Fig. 2). The reverse reaction gives back the starting cluster. A ^{1}H NMR study of the model reaction of D$_2$ with (μ-H)(μ-OPh)Os$_3$(CO)$_{10}$ indicates that the oxidative addition of H$_2$ is probably concerted with the reductive elimination of phenol since D$_2$ does not lead to a deuterated phenol (ref. 29). Here we have evidence that on a metal-metal bond interacting with a surface, an oxidative addition which is achieved at one metal atom may induce a reductive elimination at the second metal atom.

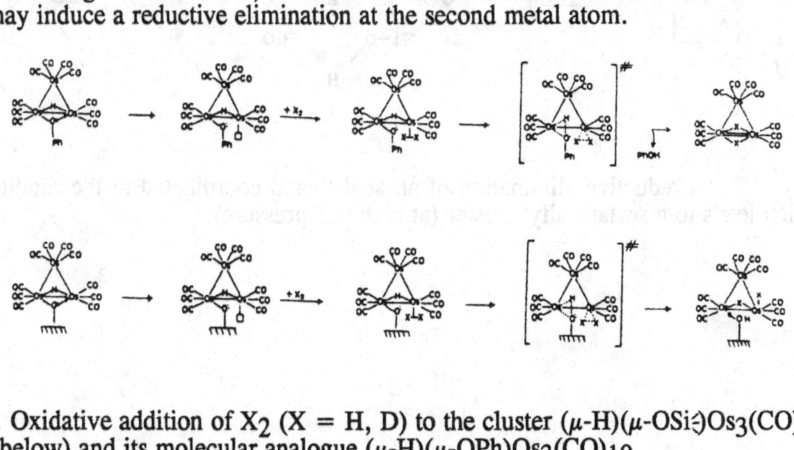

Fig. 2. Oxidative addition of X$_2$ (X = H, D) to the cluster (μ-H)(μ-OSi≤)Os$_3$(CO)$_{10}$ (below) and its molecular analogue (μ-H)(μ-OPh)Os$_3$(CO)$_{10}$

Other reactions of reversible oxidative addition of H_2 have been observed with mononuclear fragments such as $(>Si-O)(>Si-OH)Rh^I(PMe_3)_n$ ($n = 2$ or 3) (ref. 30):

$$(>Si-O)(>Si-OH)Rh(PMe_3)_n \xrightarrow{H_2} (>Si-O)(>Si-OH)Rh(H)(H)(PMe_3)_n$$

The Rh^{III} dihydride is characterized by a (Rh-H) band situated at 1938 cm^{-1}. Apparently the presence of the PMe_3 ligand is necessary to favor the oxidative addition of H_2. In the absence of PMe_3 the following equilibrium is strongly shifted to the left side (ref. 8c):

$$(>Si-O)(>Si-OH)Rh^I + H_2 \longrightarrow (>Si-O)(>Si-OH)Rh^{III}(H)(H)$$

3.4.   FORMATION OF C-C AND/OR C-O BONDS BY REDUCTIVE ELIMINATION

The surface complex $(>Si-O)(>Si-OH)Rh(\eta^3-C_3H_5)_2$ which has been obtained by reaction of $Rh(\eta^3-C_3H_5)_3$ with a partially hydroxylated silica can be considered as an 18 electron species if one applies simple rules of molecular chemistry to surface structures on oxides (refs. 2a, 30). It undergoes a series of reactions under CO which can be explained by two types of reductive eliminations:
 - a reductive elimination of hexadiene-1,5 which occurs at low CO pressure (ref. 30):

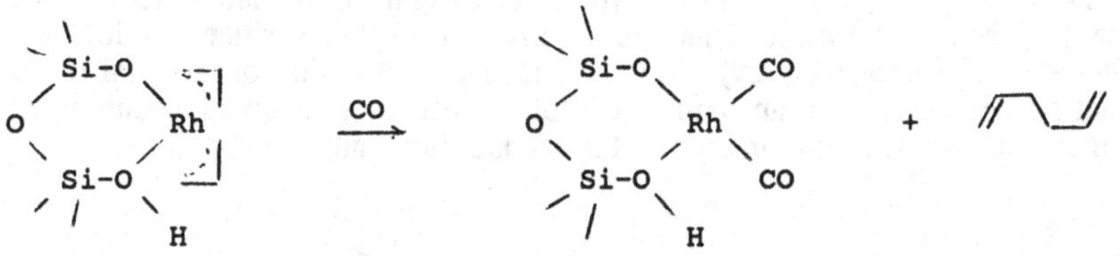

 - a reductive elimination of an acyl ligand coordinated to the rhodium(III) which leads to a surface allylic ester (at high CO pressure):

$$
\begin{array}{ccc}
\begin{array}{c}
\text{Si-O} \\
\text{O} \\
\text{Si-O} \\
\text{O} \\
\text{Si-O}
\end{array}
\quad
\overset{O}{\underset{Rh^{III}}{\overset{\parallel}{C}-CH_2}}
\overset{CH=CH_2}{}
\text{CO}
&
\longrightarrow
&
\begin{array}{c}
\text{Si-O} \\
\text{O} \\
\text{Si-O} \\
\text{O} \\
\text{Si-O}
\end{array}
\quad
\overset{O}{\overset{\parallel}{C}-CH_2}
\overset{CH=CH_2}{}
\text{Rh}\quad\text{CO}\quad\text{CO}
\end{array}
$$

This last reaction is an example where an organometallic fragment is transfered to a surface oxygen by reductive elimination (ref. 30). Although the formation of $\gg Si-O-C(O)-CH_2-CH=CH_2$ implies the insertion of CO into an allylic fragment to give an allylic acyl ligand, such an acyl ligand has not been detected. Such insertion reactions should deserve more studies in heterogeneous catalysis.

4. Surface Organometallic Chemistry as a Tool For Tailor Made Catalysts: Preparation of Bimetallic Particles From Bimetallic (or Heteropolynuclear) Clusters of Group VIII Metals

Supported bimetallic particles represent an important class of heterogeneous catalysts (ref. 32). So far, most of the methods of preparation were mainly based on co-impregnation, ion exchange or co-precipitation techniques followed by high temperature reduction. Recently heteropolynuclear molecular clusters have been used as precursors of hypothetical bimetallic particles; they exhibit in some cases unusual catalytic properties in various reactions (ref. 33), such as ethanol synthesis from syngas.

A detailed investigation of the chemistry of heteropolynuclear hydrido clusters has been undertaken in order to see whether or not these clusters could be the precursors of choice of bimetallic supported particles of predetermined composition. On a basic oxide such as partially dehydroxylated magnesia, $H_2FeOs_3(CO)_{13}$, $H_2FeRu_3(CO)_{13}$ as well as $HFeCo_3(CO)_{12}$ are deprotonated to the corresponding mono-anion (ref. 34):

$$H_2FeOs_3(CO)_{13} + \gg Mg-OH \rightarrow [HFeOs_3(CO)_3]^- (\tfrac{1}{2}Mg^{2+}) + H_2O$$

$$H_2FeRu_3(CO)_{13} + \gg Mg-OH \rightarrow [HFeRu_3(CO)_{13}]^- (\tfrac{1}{2}Mg^{2+}) + H_2O$$

$$HFeCo_3(CO)_{12} + \gg Mg-OH \rightarrow [FeCo_3(CO)_{12}]^- (\tfrac{1}{2}Mg^{2+}) + H_2O$$

The deprotonation reaction is very facile at room temperature and leads to anionic species which are well spread over the magnesia surface. After removal of the solvent under vacuum, the anionic clusters have been treated under H_2 at 400°. The resulting particles are small in size and their composition perfectly reflects that of the bimetallic cluster (ref. 35).

TABLE 1. Characterization of the bimetallic particles

Precursor complexes	Metal content/wt %	Atomic ratio Fe/M		Particle size/A	Comments
		Cluster	Particle[a]		
$H_2FeOs_3(CO)_{13}$	1.0	0.33	0.27(0.05)	10-15	(A)
$Fe_3(CO)_{12}+Os_3(CO)_{12}$	0.8	0.33	0	10-15	(B)
$H_2FeRu_3(CO)_{13}$	0.8	0.33	0.30(0.15)	10-15	(A)
$Fe_3(CO)_{12}+Ru_3(CO)_{12}$	1.1	0.33	0	10-50	(B)
$HFeCo_3(CO)_{12}$	0.9	0.33	0.33(0.03)	20-40	(A)

[a] Average ratios. The errors given in parentheses encompass all the values measured on at least 20 particles.
(A) Bimetallic particles only; (B) No Fe detected in all the particles

Co-adsorption of the two related homonuclear clusters, i.e. $(Fe_3(CO)_{12} + M_3(CO)_{12}$ (M = Ru, Os; Fe/M = 1/3) was carried out under identical conditions. In those conditions, the clusters are adsorbed as $[HFe_3(CO)_{11}]^-$, $[HRu_3(CO)_{11}]^-$ and $[HOs_3(CO)_{11}]^-$. When two of these anions are treated under H_2 as for the bimetallic clusters, one does not obtain a bimetallic particle. The reason why bimetallic molecular clusters exhibit such behavior is not clear. The temperatures at which the two monometallic clusters start to decompose into metal particles are different and this may lead to separate metallic phases. In contrast, upon destruction of the heteropolynuclear clusters, the pre-existence of metal-metal bonds might favor the formation of new metal-metal bonds between the two elements in the metal particles.

If the decomposition of the supported heteropolynuclear anions occurs under a CO atmosphere, the chemistry which occurs on the surface is different. In all cases, the presence of CO results in the formation of $Fe(CO)_5$ and anionic clusters, such as $[HOs_3(CO)_{11}]^-$, $[Os_{10}C(CO)_{24}]^{2-}$, $[HRu(CO)_{11}]^-$, or mononuclear fragments such as $\{[Co(CO)_4]^-\}_2M^{2+}$, M = Co, Mg (Scheme 1).

The fact that the surface chemistry depends quite strongly on the gaseous atmosphere present during the decomposition process is very likely related to the very easy oxidability of Fe(0) by surface OH groups even on magnesia.

5.   Surface Organometallic Chemistry on Zeolites: a New Approach For The Control of The Pore Opening Size

Shape selectivity is one of the most important properties of zeolites in their application in the field of adsorption and catalysis (ref. 36). For a given zeolite, this property is related, inter-alia, to the pore size and/or to the size of the pore openings and thus to the number of tetrahedra TO_4 that constitutes them. For a number of applications, in the field of adsorption and synthesis of fine chemicals, a given cut-off pore diameter is desired: this cannot be achieved via direct synthesis. One way to circumvent this problem is to modify, post synthesis, the pore opening of a given zeolite. This approach has been recently tried by chemical vapor deposition (CVD) of alkoxy silanes (refs. 37a-j). A final step of calcination lead to a coating of silica on the external surface of the zeolite. But although a layer by layer build up of such a coating seems to be achieved, the deposit is sometimes heterogeneous, some pores being fully blocked whereas others are not modified (ref. 37i-j).

Another way of adjusting the shape selectivity could be to chemically graft a bulky organometallic complex at the external surface of a given zeolite and more specifically and efficiently at the pore entrance. The steric hindrance of the organometallic fragment could then be easily modulated via the bulkiness of the organic moiety. This was recently achieved by reaction of bis(neopentyl)magnesium $MgNp_2$ with the external OH groups of mordenite (ref. 38):

$$\overset{\diagdown}{\underset{\diagup}{Si}}-OH \; + \quad + \; MgNp_2 \quad \rightarrow \quad \overset{\diagdown}{\underset{\diagup}{Si}}-O \diagdown \!\!\!\!\diagup Mg - \!\!\!\times \quad + \; NpH$$

The surface reaction leading to the grafted organometallic fragment is similar to the one observed under comparable conditions with silica. The amount of magnesium which can be grafted (ref. 17) never exceeds 0.2 wt %. As in the case of silica, the grafted organometallic fragment is stable up to ca. 200°C, temperature above which it starts to decompose into neopentane and an unidentified magnesium species.

Although there is no direct evidence that $Mg(Np)_2$ is grafted exclusively at the external surface of mordenite, the fact that the amount of grafted magnesium never exceeds 0.2 wt% whatever the method of preparation used (impregnation by a solution of $MgNp_2$, Et_2O or by sublimation), is in good agreement with the number of available external surface OH groups.

The adsorptive properties of these MgNp modified mordenites have been measured by a chromatographic pulse method. The adsorption capacities thus determined for n-hexane (kinetic diameter $\sigma = 4.2$ A) and isooctane ($\sigma = 6.3$ A) are shown in Fig. 3 as a function of the temperature of treatment of the Mg-mordenite. Two points should be emphasized: (i) the grafting of the organometallic moiety -Mg-Np reduces the absorption capacity of the mordenite for n-hexane by a factor of 2 and nearly suppresses the adsorption of iso-octane; (ii) as the temperature of

treatment of the modified mordenite increases, this selectivity toward n-hexane is
reduced and finally suppressed above 400°C.

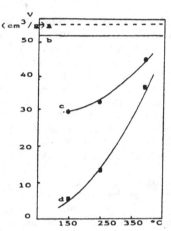

Fig. 3. Adsorption capacity of H mordenite (a, b) and MgNpH mordenite (c, d) as a
    function of temperature of treatment of the zeolite for hexane (a, c) and
    isooctane (b, d).

        This phenomenon is correlated to the thermal stability of the grafted
organometallic fragment. Computer modeling evidences different possible sites for
the grafting of MgNp entities on the external surface of mordenite (18). Among
them, a site on the (001) face, where grafting is done via two oxygen atoms (a 1e-
donor (ex silanol) and a 2e- donor, (adjacent silanols)) may be the most favorable.
No more than two such sites can exist at the entrance of a channel. When one MgNp
is grafted per pore aperture, the discrimination capacity between n-hexane and iso-
octane by MgNp modified mordenite can be easily visualized (Fig. 4):

Numerous short contact distances seem to exist between the organic molecule and the organometallic fragment in the case of isooctane (left), but none or very few in the case of n-hexane (right). When two MgNp fragments are grafted per pore aperture, no hydrocarbon can enter the pore system in agreement with the lowering of the n-hexane adsorption capacity. The results indicate that the grafting of an organometallic species on the external surface of mordenite is indeed a way of controlling shape selectivity of zeolites. It should be possible to design a shape selective system for a given application, starting from a readily available zeolite, modified by an organomtallic complex, whose steric bulk can be accurately tuned.

6.   Surface Organometallic Chemistry on Metals: New Generation of Bimetallic Catalysts Obtained by Reaction of Complexes of Main Group Elements With Group VIII Metals in Zero or Higher Oxidation State

The reaction of organometallics with supported metals (ref. 1) is still an underdevelopped area but it appears to be a very promising aspect of surface organometallic chemistry in the field of catalysis related with fine chemicals as well as petrochemicals. In fact, it appears that highly selective catalysts can be obtained by this route. So far mostly organometallics of group IV and VIII have been reacted with the surface of group VIII metals. The understanding of such chemistry is still fragmentary and requires a precise determination of the type of reaction occuring between the original supported metal and the organometallic compound as well as the precise nature of the resulting materials. At this stage of our knowledge it is premature to speak of elementary steps. We will describe here, the various reactions which occur between a supported group VIII transition metal (in high or low oxidation state) and alkyl-tin complexes. This chemistry is rather complex but some new concepts are slowly emerging. These new types of reactions lead to new and apparently well defined catalytic material that exhibit very selective properties in a variety of catalytic reactions that we shall consider later on.

6.1.   REACTION OF $Sn(n-C_4H_9)_4$ WITH Rh(0) AND Ru(0) SUPPORTED ON SILICA

When silica supported particles of zerovalent Rh or Ru are contacted with $Sn(n-C_4H_9)_4$ in the presence of hydrogen, hydrogenolysis of two Sn-C bonds occurs at a temperature as low as 25°C. The following reaction is observed (refs. 39, 40):

$$\underset{}{\overset{H}{\underset{|}{-M_s-}}} + x\ Sn(n-C_4H_9)_4 \quad \xrightarrow[\substack{0<x<1}]{\substack{25<T<100°C}} \quad M_s(Sn(n-C_4H_9)_2)_x + 2x\ C_4H_{10}$$

M_s = Rh or Ru surface atoms

The species $M_s[Sn(n-C_4H_9)_2]_x$ is a real "surface organometallic fragment" located at the surface of the metal particle:

$$
\begin{array}{ccc}
C_4H_9 & & C_4H_9 \\
\diagdown & & \diagup \\
& Sn & \\
\diagup & & \diagdown \\
Rh & & Rh
\end{array}
$$

This "surface organometallic fragment" is reasonably stable up to 200°C for Rh. Although its presence decreases to some extent the amount of CO adsorbed, it does not prevent chemisorption of CO in a linear and bridging mode:

$$
\begin{array}{ccccccc}
O & O & R & R & O & O & \\
\| & \| & \diagdown & \diagup & \| & \| & \\
C & C & Sn & C & C & \\
\diagup & \diagdown & \diagup & \diagdown & \diagup & \diagdown \\
Rh & | & Rh & Rh & | & Rh \\
& Rh & & & Rh &
\end{array}
$$

Above 100°C, the surface organometallic fragment starts to undergo a complete hydrogenolysis of the remaining Sn-C bonds and zerovalent tin atoms migrate into the particles to give an intermetallic compound:

$$
M_s[Sn(n-C_4H_9)_2]_x + xH_2 \xrightarrow{\quad 100<T<300°C \quad} M-Sn_x + 2x\ C_4H_{10}
$$

If the ratio Sn/M_s is higher than unity, slow diffusion of the tin atoms into the particle is occuring before all the alkyl-tin can react and then migrate as zerovalent atoms into the particle.

6.2. REACTION OF $Sn(n-C_4H_9)_4$ WITH Rh_2O_3/SiO_2, RuO_2/SiO_2 AND Ni_sO/SiO_2 (refs. 2b, 41)

The preparation of small particles of Rh_2O_3, RuO_2 or Ni_sO supported on silica is well known in heterogeneous catalysis. For example the preparation of Rh_2O_3/SiO_2 requires first the synthesis of small particles of supported zerovalent rhodium followed by reaction of these particles with O_2 at 100°C:

$$
2Rh^0/SiO_2 + 3/2\ O_2 \rightarrow Rh_2^{III}O_3/SiO_2
$$

In the same way RuO_2 can be prepared by reaction of zerovalent ruthenium particles with pure O_2 at $50°C$:

$$Ru^0/SiO_2 + O_2 \rightarrow Ru^{IV}O_2/SiO_2$$

With nickel, a passivated surface oxide Ni_sO can be obtained by treatment of a supported zerovalent nickel particle with a very small partial pressure of oxygen (ref. 43):

$$Ni_s/SiO_2 + O_2 \rightarrow Ni_s^{II}O/SiO_2$$

The reaction between those supported oxides and $Sn(n-C_4H_9)_4$ occurs in two steps. There is first formation of a kind of "surface organometallic complex" which is further reduced into a zerovalent intermetallic compound:

$$MO_x + zSn(n-C_4H_9)_4 \rightarrow MO_x[Sn(n-C_4H_9)_2]_2 + z\ C_4H_8 + z\ C_4H_{10}$$

$$MO_x[Sn(n-C_4H_9)_2]_2 \rightarrow MSn_z + x\ H_2O + a\ C_4H_{10} + b\ C_4H_8$$

$$(M = Ru, Rh, Ni;\ 1<z<3;\ a = 2z - x;\ b = 2z + x)$$

The overall equation is in fact a reduction of the oxide by the tetra-alkyl-tin which can be formulated very simply in the case of Rh (ref. 41), Ru (ref. 39) and Ni (ref. 43).

$$Rh_2^{III}O_3 + 4\ Sn^{IV}(n-C_4H_9)_4 \rightarrow 2\ Rh^0Sn_2^0 + 11\ C_4H_8 + 5\ C_4H_{10} + 3\ H_2O$$

$$Ru^{IV}O_2 + 2\ Sn^{IV}(n-C_4H_9)_4 \rightarrow Ru^0Sn_2^0 + 6\ C_4H_8 + 2\ C_4H_{10} + 2\ H_2O$$

$$Ni_s^{II}O + 4\ Sn^{IV}(n-C_4H_9)_4 \rightarrow Ni^0Sn_4^0 + 9\ C_4H_8 + 7\ C_4H_{10} + H_2O$$

Depending on the starting ratio SnR_4/group VIII metal, various mechanisms of alkylation, reduction and alloy formation are observed. It is not the purpose of this review article to detail these complex mechanisms. However, it must be pointed out that the composition of the final alloy or intermetallic compound is mostly determined by the amount of SnR_4 introduced at the beginning of the reaction.

6.3.   CHARACTERIZATION OF THE SUPPORTED ALLOYS

The supported alloys can be characterized by electron microscopy (ref. 39). Their particle size is usually much larger than that of the starting zerovalent particle (Fig. 5). The X-ray emission pattern indicates clearly that the particles are bimetallic (ref. 39).

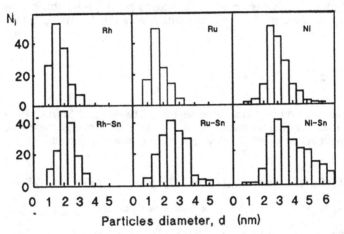

Fig. 5. Histogram of particle size distribution of various silica supported monometallic and bimetallic catalysts (Sn/Rh = 1.4; Sn/Ru = 2.7; Sn/Ni = 1.7)

In most cases, the tin signal is associated with the noble metal signal and there is no tin on the support. The EXAFS data (ref. 44), still preliminary, clearly indicate however that in the case of a Rh/Sn bimetallic particle, the average number of Rh-Rh nearest neighbour decreases considerably after alloying with tin. The XPS data (ref. 41) confirm, if necessary, the fact that rhodium and tin are in a zerovalent oxidation state (Table 2).

TABLE 2. Binding energies [eV] of $Rh(3d_{5/2}, 3d_{3/2})$ and $Sn(3d_{5/2}, 3d_{3/2})$ electrons in various rhodium, tin, and rhodium-tin samples.

Sample	Rh		Sn		Ref.
	$3d_{3/2}$	$3d_{5/2}$	$3d_{3/2}$	$3d_{5/2}$	
Rh_2O_3	313.2	308.5±0.5			[4a, b]
Rh metal	311.7	307.0±0.3			[4a-d]
SnO_2			494.9	486.4	[4a]
Sn metal			493.15	484.7	[4a]
$Rh/Sn/SiO_2$	311.6	306.9	493.5	485.0	[4e]

The use of CO as a "molecular probe" is an indirect but very powerful way of characterization of the bimetallic particles at least in the case of rhodium and nickel. It is well known that on rhodium and nickel, CO is chemisorbed both in a linear and μ_{-x} form (x is equal to 2 or 3). In the bimetallic particles the μ_{-x} form of chemisorbed CO totally disappears whereas the band corresponding to linear CO is shifted by 40-50 cm^{-1} towards low frequencies (Fig. 6). The first effect is interpreted by the concept of apparent or real "site isolation effect". The adjacent rhodium atoms which were able to coordinate CO in a μ_{-x} mode do not appear to do so in the alloy situation. The second effect might be due to either an electronic effect or to elimination of the dipole-dipole interaction between the carbonyls. More experiments are necessary to see wether or not the phenomenon is an electronic one or just a coverage effect arising from perturbation due to dipole dipole interactions.

The strong modification of the properties of this new material with respect to pure rhodium are also evidenced by the drastic modification of the chemisorption properties toward H$_2$ and CO.

Obviously the presence of tin drastically decreases the amount of CO and H$_2$ chemisorbed. The effect is more spectacular in the case of H$_2$ than in the case of CO. It is obvious that the lower number of surface rhodium atoms cannot account alone for such dramatic effect. Calorimetric data might give a partial answer to this modification of chemisorption properties.

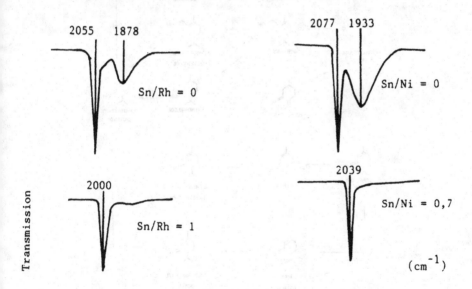

Fig. 6. Infrared spectra of CO irreversibly adsorbed at room temperature on a) Rh/SiO$_2$, b) Rh-Sn/SiO$_2$, c) Ni/SiO$_2$, d) Ni-Sn/SiO$_2$

6.4.   CATALYTIC   EFFECT   OF   AN   ORGANOMETALLIC SnR_2
        FRAGMENT   AT   THE   SURFACE   OF   A   METAL   PARTICLE:
        SELECTIVE HYDROGENATION OF CITRAL

We have seen that it is possible to modify the surface of a transition metal particle by
an organometallic fragment such as $Sn(n-C_4H_9)_2$. This surface organometallic
fragment is thermally stable under H_2 up to a temperature of 150°C. The concept of
site isolation seems to apply as well to this surface organometallic system. One may
expect, with such systems, unusual selectivities due to the presence of tin and to the
absence of ensemble of rhodium atoms.
        Citral is an interesting molecule to hydrogenate since it offers three
different types of double bonds: an isolated C=C double bond and a C=C double
bond conjuguated with a C=O double bond. Selective hydrogenation of the C=O
double bond of cis and trans citral lead respectively to geraniol and nerol which can
be hydrogenated further to give citronellol and a variety of other unsaturated and
saturated alcohols:

        The monometallic catalysts (M = Pt, Pd, Rh, Ru and Ir/SiO_2) are
unselective for the hydrogenation of citral. In contrast the same catalysts modified by
the surface organometallic fragment ($Sn(CH_3)_2$ or $Sn(n-C_4H_9)_2$) become selective
for the hydrogenation of citral (cis + trans) to geraniol and nerol (ref. 45). The

selectivity varies with the Sn/Rh ratio: at low Sn/Rh ratio the catalyst exhibits a selectivity for the hydrogenation of the conjuguated C=C double bond whereas at a Sn/Rh ratio of unity a selectivity of 96 % is observed for the carbonyl reduction. Although we have no evidence at the moment, it is quite possible that the high selectivity for the C=O bond reduction is correlated with the reaction mechanism. The possibility of complexation of the O atom of C=O to the tin(II) surface complex through a lone pair of electrons might favor the coordination of the citral by the carbonyl functionality rather than by the C=C bond:

6.5. CATALYTIC PROPERTIES OF THE BIMETALLIC ALLOYS: SELECTIVE HYDROGENOLYSIS OF ESTERS TO ACIDS OR TO ALCOHOLS

Alloys of various compositions associating either Rh, Ru or Ni with Sn can be obtained via this surface organometallic chemistry route. These new materials have a tendency to be inert towards cleavage of C-H and C-C bonds of any type of hydrocarbon. This may again be related to the fact that metal atoms in the alloy are likely isolated of their neighbours preventing the formation of the metallacycle necessary to cleave a C-C bond (ref. 45). As an example, the hydrogenolysis of propylene on a ruthenium surface is assumed to occur in the following way (ref. 45):

If the dimetallacycle of type A cannot be formed (due to the presence of tin), hydrogenolysis of the C-C bond will not occur. Preliminary experiments with Ru-Sn, Rh-Sn or Ni-Sn alloys have confirmed the fact that those systems are poor catalysts for olefin hydrogenation, olefin hydrogenolysis and homologation, as well as alkane hydrogenolysis or isomerization.

The bimetallic Rh-Sn, Ru-Sn or Ni-Sn alloys are very active and very selective for the hydrogenolysis of esters to alcohols (ref. 41). Whereas Rh metal supported on silica is able to cleave C-C and C-O bonds to give methane, ethane, CO, CO$_2$ and ethanol, Rh-Sn is able to produce ethanol with a selectivity higher than 95 % for high conversions.

7.   Conclusion

In the course of this review article, we have tried to focus on new concepts which are slowly emerging from the overlap between organometallic chemistry and surface science. It appears that some rules of molecular chemistry seem to apply quite well when organometallic complexes react with a surface, especially if the surface is that of an oxide. Well defined surface organometallic fragments can now be prepared via organometallic complexes. It is then possible to study on those fragments the real elementary steps of heterogeneous catalysis (oxidative addition, insertion, C-C coupling, a.s.o.). By means of supported heteropolynuclear complexes it is possible to prepare bimetallic particles having the same composition as that of the starting clusters, a useful approach for the synthesis of tailor made catalysts. By means of organometallics, it is possible to control the external pore size of zeolites and introduce a new approach of shape selectivation where the molecular steric control can be ajusted by the size of the organometallic fragment. Surface organometallics on metals is probably the most important from a catalytic point of view since it offers the possibility of adjusting the selectivity of supported metal catalysts by an organometallic fragment: a new generation of well defined supported bimetallic catalysts is now already used in industry.

8.   References

1.   J.M. Basset, B.C. Gates, J.P. Candy, A. Choplin, M. Leconte, F. Quignard and C. Santini (1988) "Surface organometallic chemistry: molecular approaches to surface catalysis", Kluwer Acad. Pub., The Netherlands.
       Y.I. Yermakov, B.N. Kuznetzov and V.A. Zakharov (1981), "Catalysis by supported metal complexes", Elsevier.
2a.   J.M. Basset and A. Choplin (1983), J. Mol. Catal. 21, 95.
 b.   J.M. Basset, J.P. Candy, P. Dufour, C. Santini, A. Choplin (1989), Catalysis Today 6, 1.
3.   F. Hugues, J.M. Basset, Y. Ben Taarit, A. Choplin and D. Rojas (1982), J. Am. Chem. Soc. 104, 7020.
4.   E. Guglielminotti, A. Zecchina, F. Bocuzzi and E. Borello (1980), Growth and Properties of Metal Clusters, p. 165-174, Elsevier, Amsterdam.
5a.   H.C. Foley, S.J. de Canio, K.D. Tau, K.J. Chao, J.H. Onuferko, C. Dybowski and B.C. Gates (1983), J. Am. Chem. Soc. 105, 3074.
 b.   M.D. Ward, J. Schwartz (1981), J. Mol. Catal. 11, 397.
 c.   D. Ballard (1973), Adv. Cat. 23, 263.
 d.   Y. Yermakov (1976), Catal. Rev. 13, 77.
6a.   B. Besson, B. Moraweck, J.M. Basset, R. Psaro, A. Fusi and R. Ugo (1980), J. Chem. Soc., Chem. Commun., 569-571.
 b.   R. Psaro, R. Ugo, B. Besson, A.K. Smith and J.M. Basset (1981), J. Organomet. Chem. 213, 215-247.
 c.   A.K. SMith, B. Besson, J.M. Basset, R. Psaro, A. Fusi and R. Ugo (1980), J. Organomet. Chem. 192, C31-C34.
 d.   M. Deeba and B.C. Gates (1981), J. Catal. 67, 303-307.
 e.   B. Besson, A. Choplin, L. D'Ornelas and J.M. Basset (1982), J. Chem. Soc., Chem. Commun., 843-845.

f.   J.M. Basset, B. Besson, A. Choplin and A. Théolier (1982), Philos. Trans. R. Soc. London, A308, 115-124.
g.   R. Barth, B.C. Gates, Y. Zhao, H. Knözinger and J. Hulse (1983), J. Catal. 82, 147-159.
h.   A. Choplin, M. Leconte and J.M. Basset, J. Mol. Catal. 21, 389-391.
i.   M. Deeba, B.J. Streusand, G.L. Schrader and B.C. Gates (1981), J. Catal. 69, 218-221.
j.   H. Knözinger, Y. Zhao, B. Tesche, R. Barth, R. Epstein, B.C. Gates and J.P. Scott (1982), Faraday Discuss. Chem. Soc. 72, 54-71.
k.   S.L. Cook, J. Evans and G. Neville Greaves (1983), J. Chem. Soc., Chem. Commun., 1287.
l.   S.L. Cook, J. Evans, G.S. McNulty and G. Neville Greaves (1986), J. Chem. Soc., Dalton Trans., 7-14.
m.   A. Théolier, A. Choplin, L. D'Ornelas, J.M. Basset, G. Zanderighi, R. Ugo, R. Psaro and C. Sourisseau (1983), Polyhedron 2, 119.
7.   H. Lamb and B.C. Gates (1986), J. Am. Chem. Soc. 108, 81.
8a.   A.K. Smith, F. Hugues, A. Théolier, J.M. Basset, R. Ugo, J.L. Bilhou and W.F. Graydon (1979), Inorg. Chem. 18, 3104.
b.   A. Théolier, A.K. Smith, M. Leconte, J.M. Basset, G. Zanderighi, R. Psaro and R. Ugo (1980), J. Organomet. Chem. 191, 415.
c.   J.M. Basset, A. Théolier, D. Commereuc and Y. Chauvin (1985), J. Organomet. Chem. 279, 147.
d.   J.M. Basset, B. Besson, A. Choplin and A. Théolier (1982), Philos. Trans. R. Soc. London A308, 115.
e.   H. Knözinger, E.W. Thornton and M. Wolf (1979), J. Chem. Soc. Faraday I 75, 1888.
f.   H.F. Van't Blik, J.B. Van Zon, T. Huieinga, D.C. Köningsberger and R. Prins (1983), J. Phys. Chem. 87, 2264.
g.   ibid. (1984), J. Mol. Catal. 25, 379.
h.   P. Basu, D. Panayotov, J.T. Yates Jr. (1988), J. Am. Chem. Soc. 110, 2074.
9.   R. Alves, D. Ballivet-Tkatechenko, G. Coudurier, N. Duc Chau and M. Santra (1985), Bull. Soc. Chim. Fr., 386.
10.   F. Correa, R. Nakamura, R.E. Stimson, R.L. Burwell Jr. and D.F. Shriver (1980), J. Am. Chem. Soc. 102, 5112.
11a. J.M. Basset, A. Choplin and A. Théolier, Nouv. J. Chim. 9, 654 (1985).
b.   R.A. Dalla Betta, M. Shelef (1977), J. Catal. 48, 111.
c.   V. Perrichon, M. Pijolat, M. Primet (1984), J. Mol. Catal. 25, 207.
12.   E. Guglielminotti, A. Zecchina, F. Boccuzzi and E. Borello (1980), "Growth and properties of metal clusters", 165, Elsevier, Amsterdam.
      E. Zecchina (1985), Material Chemistry and Physics 13, 379.
13.   P. Dufour, M. Leconte, C. Santini, J.M. Basset, S. Shore and S. Hsu (in press).
14.   Y. Iwasawa and H. Sato (1985), Chem. Lett., 507.
15.   F.J. Karol, G.L. Karapinka, C. Wu, A.W. Dow, R.N. Johnson and W.L. Carrick (1972), J. Polym. Sci. 10, A-1 2621.
      F.J. Karol, C. Wu, W.T. Reichle and N.J. Maraschin (1979), J. Catal. 60, 68.
16.   J.P. Candy, A. Andriollo, A. Choplin, C. Nedez and J.M. Basset (to be published).

17.   C. Chevalier, P. Ramirez de la Piscina, M. Ceruso, A. Choplin and J.M.
       Basset (1989), Catalysis Today 4, 433.
       H. Deuel, G. Huber (1951), Helv. Chim. Acta 34, 1697.
       J.J. Fripiat, J. Uytterhoeven (1962), J. Phys. Chem. 66, 800.
18.   A. Théolier, E. Custodero, A. Choplin, J. Raatz and J.M. Basset, Ang.
       Chem. Int. Ed. (in press).
19.   E.C. Ashby, G.F. Willard and A.B. Goel (1979), J. Org. Chem. 44, 1221.
20.   M.Y. He, R.L. Burwell Jr. and T.J. Marks (1983), Organometallics 2, 566.
       M.Y. He, C. Xiong, P.J. Toscano, R.L. Burwell Jr. and T.J. Marks
       (1985), J. Am. Chem. Soc. 107, 653.
21a.   A. Choplin, L. Huang, J.M. Basset, R. Mathieu, U. Siriwardane, and
       S. Shore (1986), Organometallics 5, 1547.
21b.   M.A. Drezdzon, C. Tessier-Youngs, C. Woodcockk, P. Miller Blonsky, O.
       Leal, B.K. Teo, R.L. Burwell, D. Shriver (1985), Inorg. Chem. 24, 2349.
22.   J.L. Bilhou and J.M. Basset (1977), J. Organomet. Chem. 132, 395.
23a.   W.M.H. Sachtler (1981), Proc. 8th Int. Cong. Cat. 1, 151.
  b.   S.C. Chuang, J.G. Goodwin and I. Wender (1985), J. Catal. 95, 435.
24.   H.F.S. Van't Blik, J.B. Van Zon, T. Huizinga, J.C. Vis, D.C.
       Köningsberger and R. Prins (1983), J. Phys. Chem. 87, 2264.
       H.T. Van't Blik, J.B. Van Zon, T. Huizinga, J.C. Vis, D.C.
       Koningsberger and R. Prins (1985), J. Am. Chem. Soc. 107, 3139.
       ibid (1984), J. Mol. Catal. 25, 379.
       P. Basu, D. Panayotov and J.T. Yates (1987), J. Phys. Chem. 91, 3133.
       F. Solymosi and H. Knözinger (1990), J. Chem. Soc., Faraday Trans. 86,
       389.
25.   M. Primet (1978), J. Chem. Soc., Faraday Trans., 1, 74, 2570.
26a.   W. Hieber, Sedlmeier (1954), Chem. Ber. 87, 25.
  b.   N. Homs, A. Choplin, P. Ramirez de la Piscina, L. Huang, E. Garbowski,
       R. Sanchez-Delgado and J.M. Basset (1988), Inorg. Chem. 27, 4030.
27.   R.R. Schrock, D.N. Clark, J. Sancho, J.H. Wengrovius, S.M. Rocklage
       and S.F. Peduson (1982), Organometallics 1, 1645.
28.   K. Weiss and G. Lossel (1989), Angew. Chem., Int. Ed. 28, 62.
29.   A. Choplin, L. D'Ornelas, B. Besson, R. Sanchez-Delgado and J.M. Basset
       (1988), J. Am. Chem. Soc. 110, 2783.
30.   P. Dufour, C. Santini, J.M. Basset, A. Choplin (to be published).
31.   R. Hoffman (1989), J. Am. Chem. Soc. III, 3548.
32.   J.H. Sinfelt (1973), J. Catal. 29, 308.
       J.K. Clarke, A.C. Creaner (1981), Ind. Eng. Chem., Prod. Res. Dev. 20,
       574.
33.   M. Ichikawa (1979), J. Catal. 59, 67.
34.   A. Choplin, L. Huang, J.M. Basset, R. Mathieu, U. Siriwardane and S.
       Shore (1986), Organometallics 5, 1547.
35.   A. Choplin, L. Huang, A. Théolier, P. Gallezot, J.M. Basset, U.
       Siriwardane, S. Shore and R. Mathieu (1986), J. Am. Chem. Soc. 108,
       4224.
36.   N.Y. Chen, W.E. Garwood (1986), Catal. Rev. Sci. Eng. 28, 185.
37a.   H. Olson, P.G. Rodewald (1983), (Mobil Oil Co.), US Patent 4379761.
  b.   P.G. Rodewald (1984), (Mobil Oil Co.) US Patent 4 465 886 and 4 477
       583.

c.   P.T. Allen, B.M. Drinkard, E.H. Hunger (1973), (Mobil Oil Co.), US Patent 3 724 170.

d.   I.A. Cody (1983), (EXXON Research) US Patent 4 451 572.

e.   M. Niwa, H. Ito, S. Kato, T. Hattori, Y. Murakami (1983), J. Chem. Soc., Chem. Commun., 819 and Proc. 8th Int. Cong. on Catalysis 4, 701, (1984) Berlin.

f.   M. Niwa, S. Kato, T. Hattori, Y. Murakami (1984), J. Chem. Soc., Faraday Trans. 1 80, 3135.

g.   Y. Murakami, A. Furuta, H. Ito, S. Okada (1985), Japan Patents 60 187317, 60 187318, 60 187319.

h.   M. Niwa, K. Yamazaki, Y. Murakami (1989), Chem. Lett., 441.

i.   M. Niwa, C.V. Hidalgo, T. Hattori, Y. Murakami (1986), Proc. 7th Int. Zeolite Conf., 297.

j.   M. Niwa, Y. Kawashima, T. Hibino, Y. Murakami (1988), J. Chem. Soc., Faraday Trans. 1 84, 4327.

38.   A. Théolier, A. Choplin, E. Custodero, J.M. Basset, J. Raatz, Angew. Chem. Int. Ed. (in press).

39.   M. Agnelli, P. Louessard, A. El Mansour, J.P. Candy, J.P. Bournonville and J.M. Basset (1989), Catalysis Today 6, 63.

40.   B. Didillon, J.P. Candy, P. Lesage, J.P. Bournonville and J.M. Basset (to be published).

41.   A. El Mansour, J.P. Candy, J.P. Bournonville, O.A. Ferretti, G. Mabilon and J.M. Basset (1989), Angew. Chem. 28, 347.

42.   J.P. Candy, O.A. Ferretti, G.M. Mabilon, J.P. Bournonville, A. El Mansour, J.M. Basset and G. Martino (1988), J. Catal. 112, 210.

43.   M. Agnelli, J.P. Candy, J.M. Basset, J.P. Bournonville and O.A. Ferretti (1990), J. Catal. 121, 236.

       C.H. Bartholomew and R.J. Farrauto (1976), J. Catal. 45, 41.

44.   M. Ichikawa, J.P. Candy et al., unpublished results.

45.   B. Didillon, J.P. Candy, J.P. Bournonville, A. El Mansour, J.M. Basset (1990), to be published at the International Symposium on Heterogeneous Catalysis and Fine Chemicals, Poitiers.

46.   M. Leconte, E. Rodriguez, K.I. Tanaka and K. Tanaka (1988), J. Am. Chem. Soc. 110, 275.

Index